High Energy Physics

Navin Khaneja

High Energy Physics

A Level and Transition Approach

Springer

Navin Khaneja
System and Control Engineering
Indian Institute of Technology Bombay
Mumbai, Maharashtra, India

ISBN 978-3-031-75644-3 ISBN 978-3-031-75645-0 (eBook)
https://doi.org/10.1007/978-3-031-75645-0

This Springer imprint is published by the registered company Springer Nature Switzerland AG
The registered company address is: Gewerbestrasse 11, 6330 Cham, Switzerland

Dedicated to my Parents

Preface

The book grew out of author's research in Nuclear Magnetic Resonance (NMR). Many atomic nuclei have spin, which gives it a magnetic moment, which is affected and probed by applying magnetic fields. What is the origin of the spin, the answer lies in the heart of high energy physics. The relativistic equation of electron, which respects the Einstein's mass-energy relation $E = mc^2$, is the Dirac equation, where the electron state is a four vector, tensor product of a two vector, which is the spin, with another two vector, that measures how relativistic the electron is. The four vector is called a spinor, but this is then about an electron spin. The nucleus in NMR is made of quarks, which like electrons, are Fermions, have their own spinors. But here is the problem, the nucleus is in very small femtometer, which means the kinetic energy is very high $\sim$ GeV, but mass of quarks is small with few MeV, which means quarks are very relativistic and have their spin (pointing in direction of momentum) all over the place with net spin zero. Then how do we have NMR? The answer is the beautiful interaction of electron (quark) with the vacuum. Electron can emit and then absorb a photon, raising its energy and mass (called mass correction) and making it sub GeV quark and hence non-relativistic with well-defined spin. We can think of electron (or quark), producing a field, and having energy in that field (self energy). But now, we are talking about quantum electrodynamics (QED).

Two electrons can exchange photons (one emits other absorbs) and hence exchange momentum. The net energy of a two electron system gets modified by a second-order energy correction, which is the beautiful Coulomb potential. The exchanged photon energy can further get modified, when it interacts with electron vacuum, creating electron-positron pairs and annihilating then. Modification is a second-order calculation that is the heart of high energy physics. This modification modifies the coulomb potential by a phenomenon termed vacuum polarization. An electron can directly scatter a photon, changing its and photon's momentum which is a phenomenon termed Compton scattering that is observed with a change in wavelength of a scattered X-ray light from free electrons of a crystal. Coulomb potential between moving electrons is different, which can be thought of as a part of moving frame, and due to Einstein relativity has suppressed interaction which is manifested as a magnetic field, which gives spinor its energy $\mu \cdot B$, but

μ gets modified (due to vacuum coupling) to what we call anomalous magnetic moment of electron, a second-order effect we can calculate in QED. Coupling to vacuum also modifies the orbital energies around nucleus of an atom a phenomenon that manifests itself as Lamb shift in hydrogen, a second-order effect that we can calculate in QED. Electron-Positron collide can of course scatter but can also annihilate and then create another particle-antiparticle pair, which may be heavier (like muons) as long as there is enough kinetic energy in colliding beam.

Photons are of course excitation of Electromagnetic (EM) vacuum. This is not the only vacuum; we have W-Z vacuum (weak interaction vacuum), whose excitations are heavy bosons $W+$, $W-$ and Z. Electrons can change momentum by emitting heavy boson, and changing to a neutrino, similarly quarks can change flavor by emitting, absorbing heavy bosons. Neutrinos are of course without charge but can scatter of electrons, and positrons by exchange of heavy bosons. Heavy particle can change to light particle (neutron to proton) by emitting W Boson, which can create an electron-neutrino pair by a process called β decay. Similarly pions can decay to muons and muons to electrons. These are all weak interactions because interacting boson is very heavy.

But of course, there are strong forces or color interactions that bind quarks into hadrons, mesons (pions and kaons), and Baryons (protons and neutrons). The interaction is color interaction, whereby quarks have three possible colors red, green, and blue and can exchange momentum and colors with interaction strength much higher. Color interaction is mediated by photons we call gluons, which are of eight kinds. Protons collide, exchange momentum with gluons, lose energy, and create pions and kaons, etc. Quarks are confined to protons and protons to nuclei with nuclear force. Protons are color neutral and cannot mediate interactions with exchange of gluons but can produce pions whose exchange leads to nuclear force, which is very short range as pions are massive. Gluons can interact, scatter of each other, and the exchanged gluon can change energy due to this second-order effect, which can weaken the interaction between gluons, an effect very pronounced at close distances called asymptotic freedom.

Collisions do not have to be between leptons (electron-positron, electron-neutrino), or between hadrons (proton-proton), we can collide electrons with protons. The interaction is of course EM (electro magnetic) photon. At low energies, the electron just elastically scatters of the proton and scattering cross-section dependence with exchanged momentum says how many quarks we have, three in this case. If electron is very energetic, it can excite internal modes (quark orbitals) and create a heavy proton, a process called inelastic scattering. The exchanged photon may generate pions or other mesons which is a signature of inelastic scattering.

Spinor, a four vector, is a tensor product of two vector spin and a two vector, which can be in up or down or super position. In EM, the up down states interact equally with EM vacuum, but not in weak interaction, only up state interact, a process we call parity violation, beautifully demonstrated in experiments by Lee and Yang with their collaborator Wu, in 1950s. The scattering of electron of photon doesn't depend on the gauge we use to describe the photon, we show for the same to be true for weak interactions, we need parity violation, and the mass of electron and neutrino is essentially coupling to vacuum, we call Higgs vacuum. We can excite this vacuum, creating Higg's bosons, which were detected in beautiful experiments in CERN in early 2010s. When we make a gauge transformation on heavy bosons, we have to transform Higg's field, but this should not change its energy. This is made possible if there is coupling between Higg's field and weak bosons, which give them mass and unifies the electro-weak interaction as one vacuum, a beautiful experiment performed by Salam, Weinberg, and Glashow in late 60s.

My interests in high energy were initiated by a course I took with Sidney Coleman at Harvard as a graduate student. Interest laid dormant till I began to calculate second-order corrections in NMR spectroscopy, and realized that phonon-mediated interaction in superconductivity is a second-order effect, modeled as a three level system, where two ground state levels get coupled by second-order averaging of their interaction with the excited state. This is the framework I have used to model high energy phenomenon. Emission of photon is an excitation to higher energy which returns to ground state by absorption and two phenomenon in ground state get coupled by second order averaging. Hence the name of this text, "High Energy Physics: A Level and Transition Approach." There are brief assignments after every chapter to get a good yes about the material.

I will like to thank the wonderful colleagues and academic environment of SYSCON at IIT Bombay that provided ample opportunity for self-development. Finally I would like to acknowledge the support of my family which made this effort possible.

Mumbai, India
March 2024

Navin Khaneja

Contents

Chapter 1
Historical Introduction to Particle Physics

1.1 Introduction

There are leptons (lightweight particles), the electron, muon, tau particle, and their antiparticles, like positron for electron. With their antiparticles, they also have their neutrino partners. Then, of course, there are protons and neutrons, which are hadrons. These are middleweight mesons, like pions and kaons, and heavyweight baryons, like protons and neutrons. They are made of quarks. Mesons are made of a quark-antiquark pair and baryons three quarks. Quarks also have three families, the positive ones with charge $\frac{2e}{3}$, the up (**u**) , the charm (**c**), and the top quark (**t**), and their neutrino like partners with charge $\frac{-e}{3}$, the down (**d**) , the strange (**s**), and the beauty quark (**b**), respectively. Proton is of course, uud and neutron udd. Positive Pion $\pi^+ = u\bar{d}$ and neutral pion $pi_0 = u\bar{u}$ and Kaon $K^+ = u\bar{s}$.

Protons and neutrons make the positive charged nucleus, which is surrounded by negative electrons, with electromagnetic interaction binding electron to the nucleus in an atom. In some atoms, neutron may decay into a proton giving an electron and its antineutrino, a process called β decay, mediated by what is called weak interaction. Of course, quarks are bound in a nucleus by what is called a strong interaction. And finally, we have gravitational interaction that attracts earth to the sun.

In this book, we study how different interactions between particles are mediated by different kind of photons (quanta of fields), the different excitation of the vacuum, which we can think of as big spring mass network, which oscillates as a photon (different kinds).

All of above are captured and summarized in our current knowledge of particle physics, called the standard model. Figure 1.1 shows different particles and fields in the standard model. In this chapter, we give a historical tour into how different particles and fields were discovered.

N. Khaneja, *High Energy Physics*, https://doi.org/10.1007/978-3-031-75645-0_1

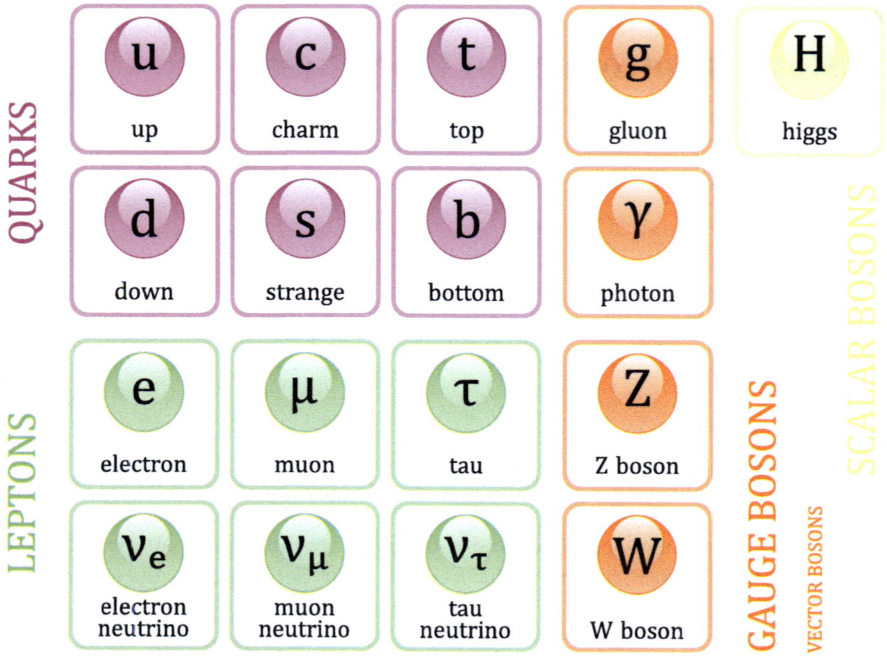

Fig. 1.1 Figure shows the main players in the standard model, the particles, and the fields

1.2 Discovery of Electron

J.J Thomson is credited with the discovery of the electron in 1897. He used a discharge tube, as shown in Fig. 1.2. The tube was filled with mercury at low pressure. When voltage was applied between electrodes, cathode rays were emitted from cathode, which could be bent using electric and magnetic fields. Direction of bend showed they were negatively charged, and Thomson coined the term electrons for them. He concluded that electrons were negative constituent of all atoms. Thomson knew atom was neutral, so there must be positive charge in the atom, and he postulated the plum pudding model, where electrons are plums in the pudding of positive charge. The model was later disproved by Thomson gold foil experiment that showed positive charge is not a pudding but concentrated in center of the atom (Fig. 1.3).

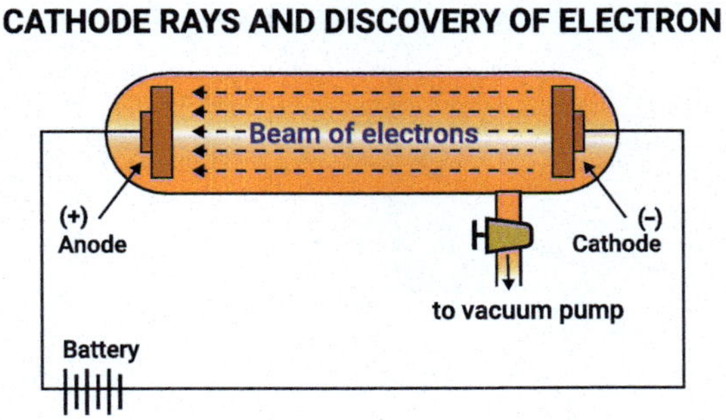

Fig. 1.2 Figure shows the apparatus used by J.J. Thomson for discovery of electron

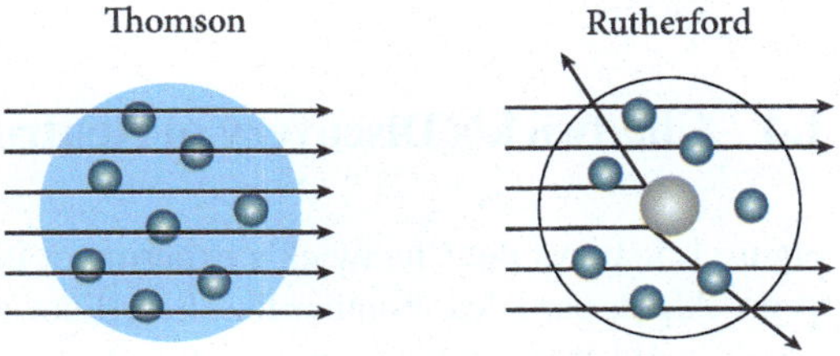

Fig. 1.3 Figure shows the plum pudding model of J.J. Thomson, where electrons are plums in a positively charged pudding

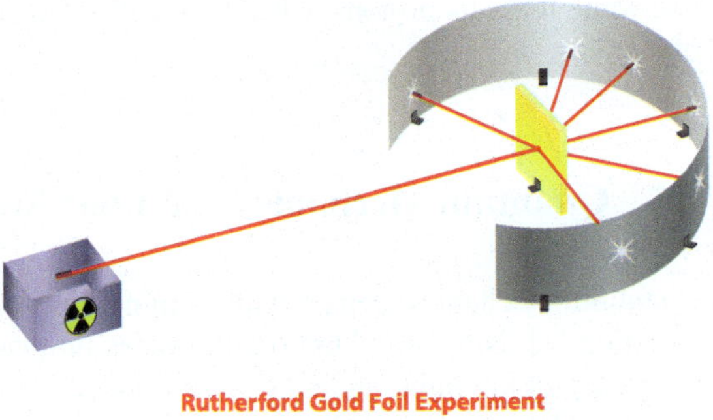

Fig. 1.4 Figure shows the gold foil experiment of Rutherford

1.3 Rutherford Gold Foil Experiment and the Proton

In 1909, Ernest Rutherford carried the following experiment. He fired alpha particles (helium nuclei) at a thin gold foil and saw the scattering of alpha particles. Most of the particles went through, while few were deflected by large angles. This suggested positive charge was concentrated into the center, called nucleus of atom. Nucleus with one unit of positive charge is a proton. This changed the model of atom from Thomson's plum pudding to Rutherford model, where positive charge is in the center of the atom, with electrons surrounding it. Nucleus is not all protons; it can have neutral particles with same mass as protons called neutrons. They were discovered by Chadwick (Fig. 1.4).

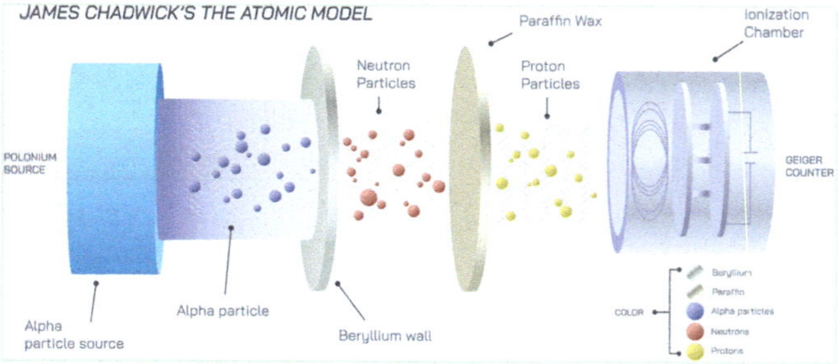

Fig. 1.5 Figure shows the Chadwick's experiment for discovery of neutron

1.4 Chadwick's Discovery of Neutron

Figure 1.5 shows the Chadwick's experiment in 1932 that led to discovery of neutrons. Alpha particles, bombarding beryllium target, emitted chargeless radiation, which could transfer all its momentum to protons in paraffin wax, meaning they had same mass as protons. Chadwick called them neutrons, important component of nucleus.

1.5 Quantum Mechanics and Bohr Model of Atom

Quantum mechanics began with seminal work of Max Planck on blackbody radiation [1]. Number of light field modes in a cavity grow $\propto \omega^2$, where ω is the frequency of the light mode. With each mode carrying kT amount of energy, there is ultraviolet catastrophe. By assuming, mode energy comes in quanta of size, $\hbar\omega$, we freeze the mode, when $\hbar\omega > kT$, preventing the catastrophe. This is shown in Fig. 1.6.

The idea of quantized energy was adopted by Einstein [2], in explaining photoelectric effect, where not light intensity, but frequency, determines if an electron is ejected, by making sure $\hbar\omega > W$ where W is the work function of the electron in solid (Fig. 1.7). Louis De Broglie [3] extended the idea of quantum, to electron, saying electron is a wave with energy $\hbar\omega$ and showed that it has momentum $\hbar k$. Electron scattering experiments by Davisson and Germer, of a nickel crystal, showed electron indeed had wave properties [4].

Bohr [5] used the momentum of electron wave to quantize the angular momentum of orbitals to integral multiples of $\hbar$, deriving energy of stationary orbits of an atom. This was the birth of quantum theory of atom and quantum mechanics (Fig. 1.8).

The quantum mechanics is developed in the hands of Heisenberg [6] and Erwin Schrödinger [7–10] as matrix mechanics and wave mechanics, respectively.

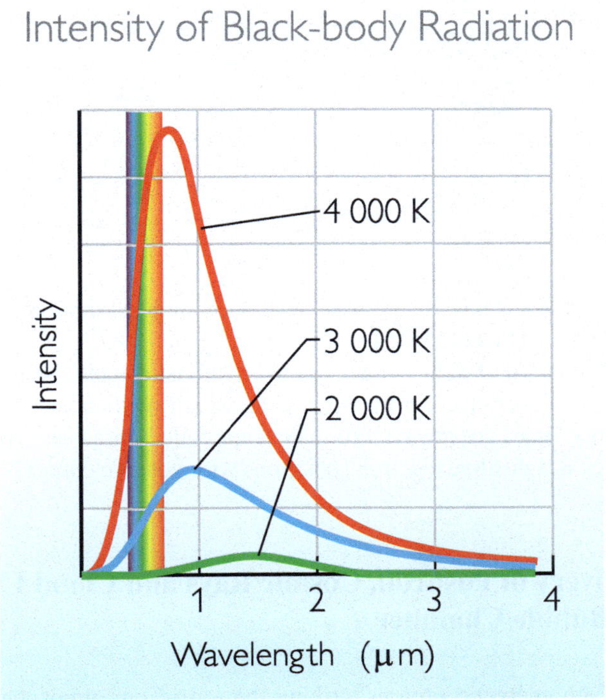

Fig. 1.6 Figure shows the blackbody radiation, where power in higher wavelengths diminishes

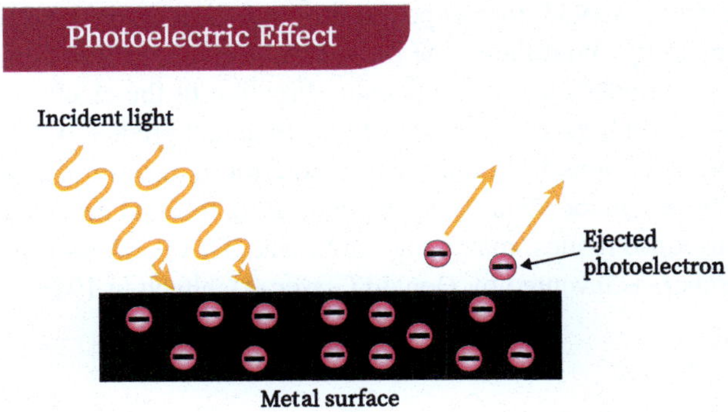

Fig. 1.7 Figure shows the photoelectric effect

Fig. 1.8 Figure shows the Bohr model of atom

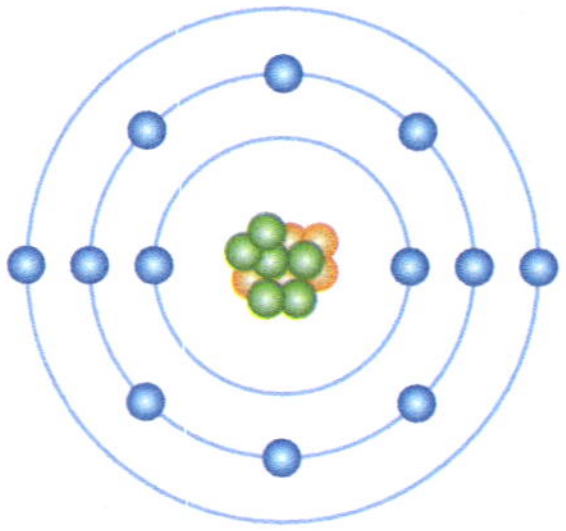

Quantum mechanics is now a very well-developed subject with numerous excellent expositions on the subject [11–15].

The wave equation of Schrödinger is a nonrelativistic approximation to the fully relativistic equation of electron by Dirac [11], which predicts electron waves with negative energy states are always full, a concept called Dirac sea. Any vacancy or hole in this sea is a positive electron (positron) with positive energy.

1.6 Discovery of Positron, Cosmic Rays and Cloud Chamber, and Bubble Chamber

Cosmic rays are energetic protons striking the top of our atmosphere, producing pions which decay to muons and positrons. These travel down and can be detected on earth. They can ionize a gas of atoms. The charged ionized atom is a site on which moisture can condense, to give a track that is seen by a camera, the one used by Anderson, in 1933, to find positrons. How these tracks curved in magnetic field, told something about their e/m ratio, and for positron was found identical to the electron. The tracks curved in opposite direction to the electron, saying it has positive charge and is exactly the antiparticle Dirac had predicted (Fig. 1.9).

We can ionize protons in liquid form, giving more density of atoms, and more vivid tracks. Once you ionize protons, in liquid form, if liquid has dissolved bubbles, they will go to ionized sites, making a vivid track. This is the working principle of a bubble chamber, as invented by Donald Glaser at Caltech in 1952.

1.7 Discovery of Muon

Muons are leptons, like electrons, just 200 times heavier. They are in cosmic showers and were recorded on a photographic emulsion by American physicists Carl D. Anderson and Seth Neddermeyer in 1936. Emulsions are also ionized by energetic particles, and the tracks are just blackened paths on the photographs. The muon discovery was confirmed in 1937 by J. C. Street and E. C. Stevenson in a cloud

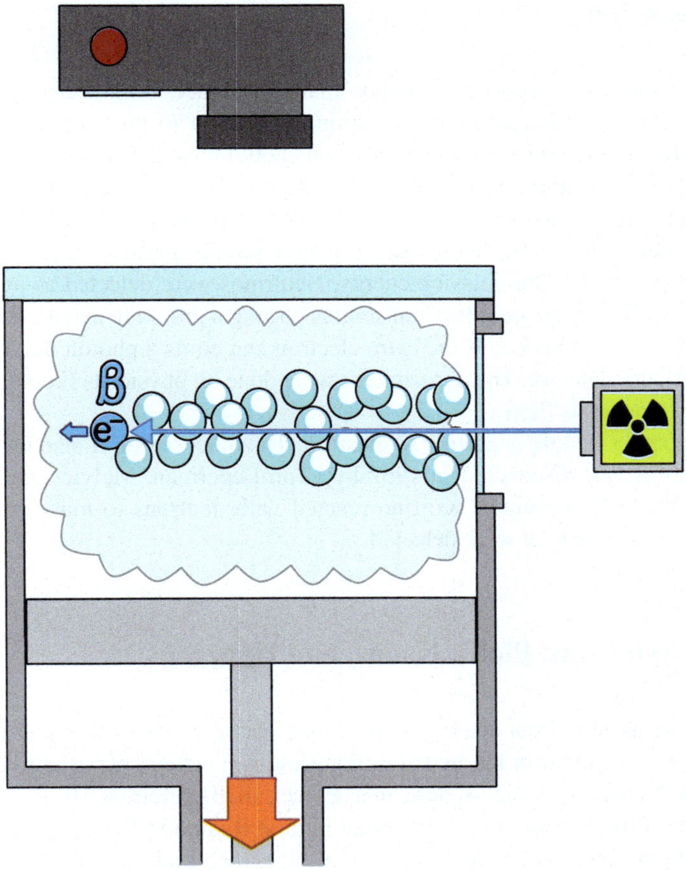

Fig. 1.9 Figure shows a typical Wilson cloud chamber

chamber. At that point, we were looking for a subatomic particle of around this mass, because theoretical physicist Hideki Yukawa has had argued that the strong force that bind protons and neutrons in a nucleus must be due to exchange of a particle of this mass so that it is short range. When muon was discovered, it was thought that Yukawa particle has been discovered, but later it turned out Yukawa particle was actually a pion discovered later.

1.8 Photographic Emulsions and Discovery of Pions

Pion is a subatomic particle found in cosmic showers that was recorded on a emulsion by group of Cecil Powell in 1947. It had same mass as predicted by Yukawa and was immediately accepted as carrier of nuclear force.

1.9 Neutrino

There is a very interesting phenomenon that takes place in nuclear physics called β decay. Many atoms transmute by turning a neutron to proton and emitting an electron. It was thought the emitted electron should have the same energy dictated by energy loss in going from neutron to proton, but what was found was that electron energy had a spectrum. This was a problem which was solved by Pauli, who postulated that along with the electron a second particle called neutrino is emitted which carries the missing energy. Neutrinos were detected as antineutrino that are absorbed by an atom that transmutes turning a proton to a neutron and emits positron. The positron decays the extra electron and emits a photon that is detected by a scintillator detector. The experiments were done by physicists Frederick Reines and Clyde Cowan in 1956.

The neutrino we talk is partner of electron. The other lepton μ also has a partner μ neutrino. It was discovered in 1962 by Leon Lederman, Melvin Schwartz, and Jack Steinberger. The muon neutrino reacted with neutrons to make protons and gave away muons which were detected.

1.10 Cyclotrons, Pions, Kaons, and Hyperons

Cosmic rays are excellent source of subatomic particles, pions and muons. But we can also produce them in lab by smashing protons. For this, we need to accelerate them to high energies, which is done in a device called cyclotron. The cyclotron was invented by Ernest Lawrence in 1929 and built at Radiation Lab Berkeley in 1931. A magnetic field keeps electrons in a spiral trajectory, and they are accelerated by a time-varying electric field. Time-varying electric field acceleration was preceded by electrostatic acceleration in device called Van de Graaff generator which was a particle accelerator.

Once protons are accelerated to high energies, they can bombard a liquid hydrogen target in a bubble chamber, and we can see wealth of new particles pions, kaons, and hyperons like Λ, Σ, whose tracts are disambiguated in bubble chamber pictures by magnetic fields around which these particles spiral (Fig. 1.10).

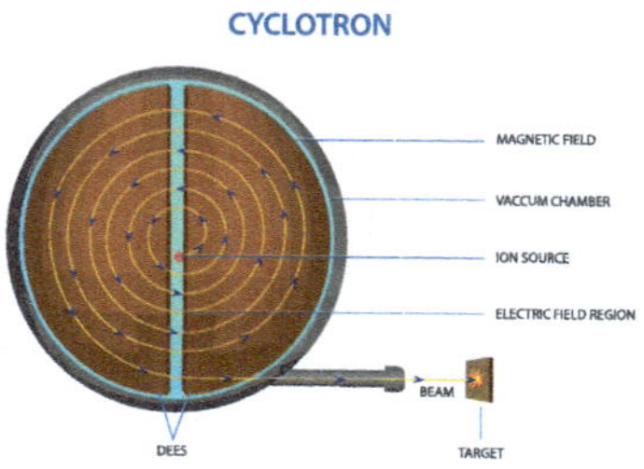

Fig. 1.10 Figure shows schematic of a cyclotron

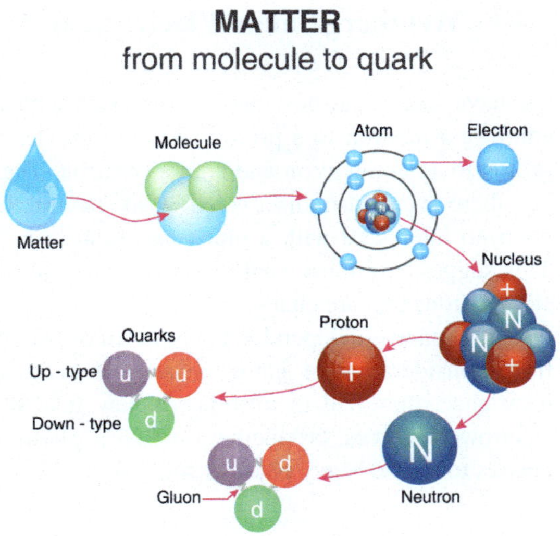

Fig. 1.11 Figure shows matter made of quarks

1.11 Quark Model

After many subatomic particles were discovered, there was a need for a model. Such a model was provided by Murray Gell-Mann in 1964. The quarks were postulated as the basic constituents of hadrons as shown in Fig. 1.11. Mesons are made of a quark-antiquark pair and baryons three quarks. Quarks also have three families, the positive ones with charge $\frac{2e}{3}$, the up (**u**) , the charm (**c**), and the top quark (**t**), and their neutrino like partners, with charge $\frac{-e}{3}$, the down (**d**) , the strange (**s**), and the beauty quark (**b**), respectively. Proton is of course, uud and neutron udd. Positive Pion $\pi^+ = u\bar{d}$ and neutral pion $pi_0 = u\bar{u}$ and Kaon $K^+ = u\bar{s}$.

Gell-Man knew only three quark u,d,s and explained known mesons and baryons in terms of these by organizing them as octets (spin 1/2) and decuplets (spin 3/2).

1.12 More Quark Flavors

More evidence for quark model came in the 1970s, when high-energy collision of electron-positron (Ritcher's group) and protons (Ting's group) produced heavy mesons ψ meson or charmonium, which was widely recognized as a bound pair of charm quark and antiquark. Even heavier quarks (**b** beauty) were found in Fermilab in 1977, by group of Leo Ledermann in proton-proton collisions, that created bottomium. Heaviest quark the top quark was found in Fermilab experiments in 1995.

1.13 Weinberg Salam Electroweak Model

We have seen neutrinos; they can interact with matter, like an electron; neutrino can turn a neutron to a proton and emit an electron or an antineutrino can turn a proton to an antineutron and emit a positron. These are charged weak interactions, but there were also discovered weak neutral interactions whereby an electron neutrino interacted with a muon and scattered. The scattering event was neutral. This suggested a third weak neutral photon, the Z photon, in addition to W^+ and W^- photons that are charged.

Z photon was independently predicted by Weinberg, Salam, and Glasgow (1968), from consideration of gauge invariance of weak interactions. To obtain gauge invariance, they had to postulate a new field the Higgs field, which unifies the electroweak forces, producing a massless photon and a heavy Z boson, and gives masses to all the heavy photons and fermions.

1.14 Discovery of W-Z Bosons

The W-Z boson predicted by Weinberg, Salam, and Glasgow (1968) were discovered in CERN lab in (1983) by group of Carlo Rubbia. They collided protons (uud) and antiprotons ($\bar{u}\bar{u}\bar{d}$) to give $u\bar{d} \to W^+$ and $\bar{u}d \to W^-$ bosons, which decayed to positrons and electrons, respectively, which were detected. Masses were estimated to around 80 GeV. The $u\bar{u} \to Z$ was also detected by its decay to electron-positron pair, with mass estimated to 90 GeV. Clearly, the center of mass energy for these collision measurements is around $> 90 \times 3 \sim 270$ GeV.

1.15 Discovery of Higgs Boson

Higgs boson was discovered at CERN in 2013, in two big projects ATLAS and CIMMS. Protons at Teravolt energies collide to generate top quark-antiquark pairs, which recombine to give Higgs boson as in Fig. 1.12. Higgs couple to W and Z bosons, which are produced subsequently, and decay in electrons and positrons that are detected.

1.16 Colliders and Detectors

If we collide two particles, head on, at relativistic energies, then we can ask how is it in a frame, when one is at rest, then the other particle has an energy many folds, which means it is not efficient to not collide particles head on in center of

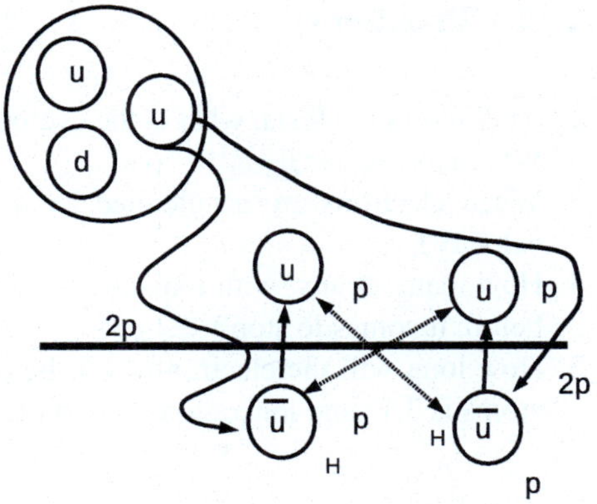

Fig. 1.12 Figure shows Higgs production from proton collisions

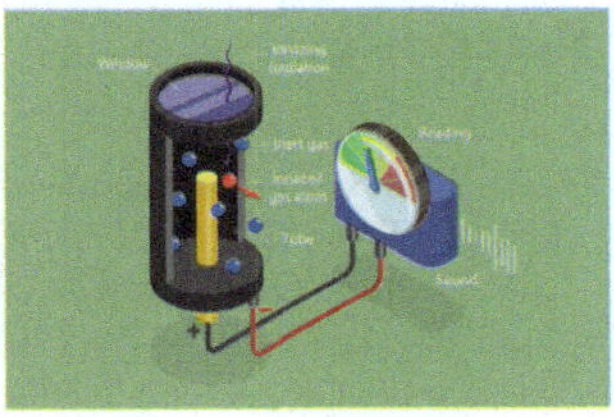

Fig. 1.13 Figure shows working of a Geiger counter

mass (CM) frame. Thus, initial particle accelerators, which were cyclotrons and synchrotrons that accelerated particles and collided them with a stationary target, are being replaced with colliders that do head on. Colliders are electron-positron colliders like LEP at CERN, which can go up to 100 GeV energies or proton-proton colliders like LHC at CERN which can of course be very high 10s of TEV energies.

Collisions give charged particles whose tracks can be detected by 3D frame of thin wires who get capacitively coupled with charge accumulating in the (x, y) coordinate where particle passes by. This is a wire chamber. These are of course modern versions of cloud and bubble chambers we have talked about. Particle detectors mainly work on principle of ionization. The most rudimentary ones like Geiger counters detect ionizing radiation or particles by ionizing atoms in a chamber and detecting emitted electrons. Ionizing radiation or high-energy photons can be detected by scintillators which promote valence electrons to conduction electrons by a second-order Raman process. These then emit photons, which are read on a photodiode. Energy of high-energy electrons, positrons can be read in a colorimeter, whose temperature rises when it stops a fast-moving electron. Muons do not lose energy to the medium as much as electrons because their mass is comparable to atomic mass. They pass through many layers of calorimeter when finally they are detected by say scintillation. Neutrinos (antineutrino in fact) of course interact with matter turning a proton to neutron or vice versa, and resulting positron can annihilate electrons to give high-energy photon detected by scintillators (Fig. 1.13).

1.17 Problems

1. At B field of 1 Tesla, what is the cyclotron frequency of a 1 GeV electron?
2. What is its radius?
3. When electrons are accelerated by a field of 10kV, from rest, what is their velocity?
4. How many atoms (with ionization energy 10 eV) will electron at 1 GeV ionize before it comes to stop?
5. How long will the electron travel, before it comes to stop, if distance between atoms is $3A^{\circ}$ and ionization probability is 10^{-3}?

References

1. M. Planck, On the theory of the energy distribution law of the normal spectrum. Verhandl. Dtsch. Phys. Ges. **2**, 237 (1900)
2. A. Einstein, On the heuristic point of view of production and transformation of light. Annalen der Physik **17**, 132–148 (1905)
3. L. De Broglie, A tentative theory of light quanta. Phil. Mag. **47**, 446–458 (1924)
4. C. Davisson, L.H. Germer, Diffraction of electrons by a crystal of nickel. Phys. Rev. **30**, 705 (1927)
5. N. Bohr, *The Theory of Spectra and Atomic Constitution; Three Essays* (Cambridge University Press, Cambridge, 1922)
6. W. Heisenberg, Uber quantentheoretische Umdeutung kinematischer und mechanischer Beziehungen. Z. Phys. **33**, 879–893 (1925)
7. E. Schrödinger, Quantisation as a problem of proper values (part I). Annalen der Physik (4) **79**, 1–7 (1926)
8. E. Schrödinger, Quantisation as a problem of proper values (part II). Annalen der Physik (4) **79**, 102–118 (1926)
9. E. Schrödinger, On the relation between the quantum mechanics of Heisenberg, Born, and Jordan, and that of Schrodinger. Annalen der Physik (4) **79**, 361–376 (1926)
10. E. Schrödinger, The continuous transition from micro- to macro-mechanics. Die Naturwissenschaften **28**, 664–666 (1926)
11. P.A.M. Dirac, The fundamental equations of quantum mechanics. Proc. R. Soc. Lond. Ser. A **109**, 642–653 (1926)
12. C. Cohen-Tannoudji, B. Diu, F. Laloe, *Quantum Mechanics*, vols. 1 and 2, 1st edn. (Wiley, 1991)
13. J.J. Sakurai, J. Napolitano, *Modern Quantum Mechanics*, 3rd edn. (Cambridge University Press, 2020)
14. D.J. Griffiths, *Introduction to Quantum Mechanics* (Pearson Prentice Hall, 2004)
15. D. Griffiths, *Introduction to Elementary Particles* (Wiley-VCH, 2004)

Chapter 2
Relativity, Electrons, and Photons

2.1 Notation

Three vectors are denoted by boldface type.

$$x^\mu = (x^0, \mathbf{x}) \tag{2.1}$$

$$x_\mu = (x^0, -\mathbf{x}) \tag{2.2}$$

$$\partial_\mu = (\frac{\partial}{\partial x^0}, \boldsymbol{\nabla}) \tag{2.3}$$

$$\partial^\mu = (\frac{\partial}{\partial x^0}, -\boldsymbol{\nabla}) \tag{2.4}$$

$$x^\mu x_\mu = (x^0)^2 - (x^1)^2 - (x^2)^2 - (x^3)^2$$

and

$$\partial_\mu \partial^\mu = (\frac{\partial}{\partial x^0})^2 - (\frac{\partial}{\partial x^1})^2 - (\frac{\partial}{\partial x^2})^2 - (\frac{\partial}{\partial x^3})^2$$

2.2 Relativity

Consider lab frame O and a frame O' , moving with respect to lab frame with velocity v in the x-direction as shown in Fig. 2.1.

Then, the space-time increment $(\Delta x, \Delta t)$ in O corresponds to $(\Delta x', \Delta t')$ in O'. The phase increment of the light wave in both frames is the same. The velocity of light is same in both frames, which is the central tenet of theory of relativity.

N. Khaneja, *High Energy Physics*, https://doi.org/10.1007/978-3-031-75645-0_2

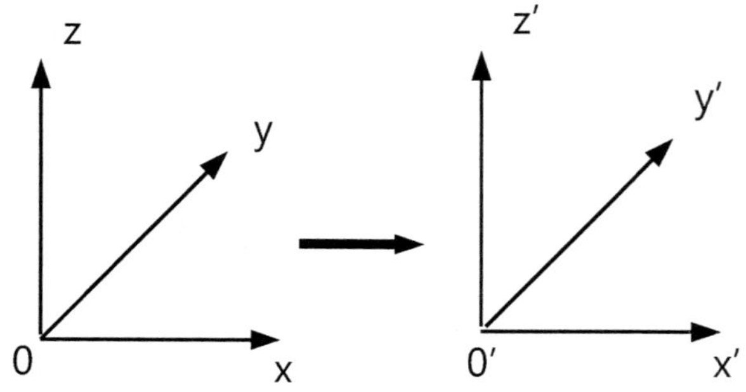

Fig. 2.1 Figure shows frames O' moving relative to O at velocity v

Then

$$k\Delta x - \omega\Delta t = k'\Delta x' - \omega'\Delta t' \tag{2.5}$$

$$k(\Delta x - c\Delta t) = k'(\Delta x' - c\Delta t'). \tag{2.6}$$

For light travelling in opposite direction

$$k'(\Delta x + c\Delta t) = k(\Delta x' + c\Delta t'). \tag{2.7}$$

The two relations give

$$(c\Delta t)^2 - \Delta x^2 = (c\Delta t')^2 - \Delta x'^2. \tag{2.8}$$

For $\Delta x' = 0$, we have $\Delta x = v\Delta t$, and this gives

$$\Delta t = \frac{\Delta t'}{\sqrt{1 - \frac{v^2}{c^2}}} \tag{2.9}$$

This is called time dilation. Furthermore

$$\frac{k'}{k} = \frac{1 - \frac{v}{c}}{\sqrt{1 - \frac{v^2}{c^2}}} \tag{2.10}$$

Then, combining Eqs. 2.6, 2.7, 2.10, we get

$$\begin{bmatrix} \Delta x \\ c\Delta t \end{bmatrix} = \frac{1}{\sqrt{1 - \frac{v^2}{c^2}}} \begin{bmatrix} 1 & \frac{v}{c} \\ \frac{v}{c} & 1 \end{bmatrix} \begin{bmatrix} \Delta x' \\ c\Delta t' \end{bmatrix} \tag{2.11}$$

For a rod of length l' in O' we have $(\Delta x', \Delta t') = (l', 0)$, the $l = \Delta x - v\Delta t = l'\sqrt{1 - \frac{v^2}{c^2}}$. This is called length contraction.

For an object moving at velocity in the frame O' at velocity u, for time $\Delta t'$, we have $(\Delta x', \Delta t') = (u\Delta t', \Delta t')$. Then, from (Eq. 2.11), the relative velocity

$$v = \frac{\Delta x}{\Delta t} = \frac{u + v}{1 + \frac{uv}{c^2}} \tag{2.12}$$

Of course, the world is three dimensional, with $X = (x, y, z, ct)$, we have for O moving with v along x direction to O', we have for $\gamma = \sqrt{1 - \frac{v^2}{c^2}}$

$$\Delta X = \begin{pmatrix} \frac{1}{\gamma} & \frac{v}{c\gamma} & 0 & 0 \\ \frac{v}{c\gamma} & \frac{1}{\gamma} & 0 & 0 \\ 0 & 0 & 1 & 0 \\ 0 & 0 & 0 & 1 \end{pmatrix} \Delta X' \tag{2.13}$$

Now, with $u = (u_x, u_y, u_z)$, we have

$$v = (\frac{u_x + v}{1 + \frac{u_x v}{c^2}}, \frac{u_y \gamma}{1 + \frac{u_x v}{c^2}}, \frac{u_z \gamma}{1 + \frac{u_x v}{c^2}}) \tag{2.14}$$

For conservation of momentum to hold in relativistic frame transformation, we have to define momentum as

$$\mathbf{p} = \frac{m}{\gamma}(v_x, v_y, v_z) \tag{2.15}$$

Kinetic energy can be computed by finding work done in accelerating m_0 from rest to v; this is

$$W = \int d(mv)v = \int d(mv^2) - mvdv = mv^2 + \int d(m_0 c^2 \sqrt{1 - \frac{v^2}{c^2}} = mc^2 - m_0 c^2 \tag{2.16}$$

But we have rest mass energy $m_0 c^2$, giving total energy mc^2. To see rest mass energy, let the energy of the rest mass m_0 in O' be U. Then, its energy in O is $U + (m - m_0)c^2$. Let this mass disintegrate giving two photons in forward and backward direction of energy $\hbar\omega_0 = \frac{U}{2}$ each. Then, in frame 0, the energies of photons are $\hbar\omega_1$ and $\hbar\omega_2$. Then, we get

$$\hbar\omega_1 + \hbar\omega_2 = 2\hbar\omega_0(\frac{\omega_1}{2\omega_0} + \frac{\omega_2}{2\omega_0}) = \frac{2\hbar\omega_0}{\sqrt{1 - \frac{v^2}{c^2}}}$$

This gives $U = m_0c^2$ and $E = mc^2$.

We can define $p = c(mc, p_x, p_y, p_z)$, and then, a direct verification gives the frame transformation:

$$p = \begin{pmatrix} \frac{1}{\gamma} & \frac{v}{c\gamma} & 0 & 0 \\ \frac{v}{c\gamma} & \frac{1}{\gamma} & 0 & 0 \\ 0 & 0 & 1 & 0 \\ 0 & 0 & 0 & 1 \end{pmatrix} p' \tag{2.17}$$

Consider now mass m_0 moving with velocity $u = (u_x, u_y, u_z)$, its energy $E = \frac{m_0c^2}{\sqrt{1-\frac{u^2}{c^2}}}$ increments slightly if we go in the frame moving with velocity v such that the new velocity is given by Eq. 2.14, and

$$\frac{dE}{dv_x}|v_x = 0 = mu_x,$$

this is a new interpretation of x momentum, and it is differential energy change as we move in a frame in x direction. Similarly, y and z momentum can be defined. Now, let us use this new interpretation of momentum.

Consider a electron matter wave with frequency, wave vector (ω, k) and (ω', k'), respectively.

The phase increment of the matter wave in both frames is the same.

Then

$$k\Delta x - \omega\Delta t = k'\Delta x' - \omega''\Delta t' \tag{2.18}$$

$$\begin{bmatrix} k & -\frac{\omega}{c} \end{bmatrix}\begin{bmatrix} \Delta x \\ c\Delta t \end{bmatrix} = \begin{bmatrix} k & -\frac{\omega}{c} \end{bmatrix}\frac{1}{\sqrt{1-\frac{v^2}{c^2}}}\begin{bmatrix} 1 & \frac{v}{c} \\ \frac{v}{c} & 1 \end{bmatrix}\begin{bmatrix} \Delta x' \\ c\Delta t' \end{bmatrix} = \begin{bmatrix} k' & -\frac{\omega'}{c} \end{bmatrix}\begin{bmatrix} \Delta x' \\ c\Delta t' \end{bmatrix} \tag{2.19}$$

This gives

$$\begin{bmatrix} k & -\frac{\omega}{c} \end{bmatrix}\frac{1}{\sqrt{1-\frac{v^2}{c^2}}}\begin{bmatrix} 1 & \frac{v}{c} \\ \frac{v}{c} & 1 \end{bmatrix} = \begin{bmatrix} k' & -\frac{\omega'}{c} \end{bmatrix} \tag{2.20}$$

Rewriting this equation, we get

$$\begin{bmatrix} k \\ \frac{\omega}{c} \end{bmatrix} = \frac{1}{\sqrt{1-\frac{v^2}{c^2}}}\begin{bmatrix} 1 & \frac{v}{c} \\ \frac{v}{c} & 1 \end{bmatrix}\begin{bmatrix} k' \\ \frac{\omega'}{c} \end{bmatrix} \tag{2.21}$$

Once again, we use our interpretation of momentum and ask what is $\frac{dE(v)}{dv}|_0 = \hbar k'$. Therefore, momentum of our complex wave ω', k' is simply

$$\hbar k'.$$

Thus, we have two basic results in quantum mechanics; the energy is $\hbar\omega$ and momentum $\hbar k$.

Also note that directly by definition $E = \sqrt{(pc)^2 + (m_0 c^2)^2}$.

2.3 Photon

Consider Maxwell's equations in free space in coordinate system O.

$$\nabla \cdot E = 0, \tag{2.22}$$

$$\nabla \cdot B = 0, \tag{2.23}$$

$$\nabla \times B = \mu_0 \frac{\partial D}{\partial t}, \tag{2.24}$$

$$\nabla \times E = -\frac{\partial B}{\partial t}. \tag{2.25}$$

In coordinate system O', the E, B transform to E', B' such that Maxwell's equations stay same, i.e., the new E' and B' fields should also satisfy Maxwell's equations. So, what should be the transformation rule. Recall, we call write from 2.23 that

$$B = \nabla \times \mathbf{A}. \tag{2.26}$$

($\mathbf{A} = (A_x, A_y, A_z)$) and substituting in 2.25, we get

$$E = -\frac{\partial \mathbf{A}}{\partial t} - \nabla A_0. \tag{2.27}$$

With $A^\mu = (A_0, \mathbf{A})$, observe the gauge transformation

$$A_\mu \rightarrow A_\mu - \partial_\mu \chi, \tag{2.28}$$

does not change E and B so we choose Lorentz gauge

$$\partial_\mu A^\mu = 0. \tag{2.29}$$

Substituting for E, B in terms of A in Eq. 2.22 and 2.24, we find

$$\partial^{\mu}\partial_{u}A^{\nu} = 0. \tag{2.30}$$

Now, define

$$A'(x'(x)) = \Lambda\ A(x). \tag{2.31}$$

Then, we can check that

$$\partial_{\mu}A'^{\mu} = 0, \tag{2.32}$$

and

$$\partial^{\mu}\partial_{u}A'^{\nu} = 0. \tag{2.33}$$

Now, we can define E' and B' in terms of A' as in 2.26 and 2.27; this ensures that B' and E' satisfy Maxwell's equation 2.23 and 2.25. Then, using 2.32 and 2.33, we get E' and B' also satisfy 2.22 and 2.24. Therefore, the new E' and B' fields also satisfy Maxwell's equations. Remember, transformation rule is 2.31.

Electromagnetic field is

$$F^{\mu\nu} = \partial^{\mu}A^{\nu} - \partial^{\nu}A^{\mu} = \begin{bmatrix} 0 & -E_x & -E_y & -E_z \\ E_x & 0 & -B_z & B_y \\ E_y & B_z & 0 & -B_x \\ E_z & -B_y & B_x & 0 \end{bmatrix} \tag{2.34}$$

We now give a variational interpretation to the equation of A field 2.33. We equip A with a dynamics by defining Lagrangian as density

$$L = -\frac{\epsilon_0}{4}F_{\mu\nu}F^{\mu\nu} = \frac{\epsilon_0}{2}(E^2 - B^2). \tag{2.35}$$

The corresponding energy density is

$$H = \epsilon_0(F_{0\mu}F^{0\mu} + \frac{1}{4}F_{\mu\nu}F^{\mu\nu}) = \frac{\epsilon_0}{2}(E^2 + B^2). \tag{2.36}$$

Once we have the Lagrangian, we can write the Euler Lagrange equations that give us Eq. 2.33. Let's make a small detour on how Euler Lagrange equations arise from Lagrangian.

2.3.1 Euler Lagrange Equations

Recall given a mechanical system with Lagrangian $L(x, \dot{x})$, we want to find the trajectory connecting two fixed points that minimize

$$S = \int L(x, \dot{x})\, dt. \tag{2.37}$$

$$\delta S = \int \delta L(x, \dot{x})\, dt, \tag{2.38}$$

$$\delta S = \int \frac{\partial L}{\partial x}\delta x + \frac{\partial L}{\partial \dot{x}}\delta \dot{x}. \tag{2.39}$$

Integrating by parts with δx as 0 at the end points/boundary, we have

$$\delta S = \int (\frac{\partial L}{\partial x} - \frac{d}{dt}\frac{\partial L}{\partial \dot{x}})\delta x. \tag{2.40}$$

For above to be true for arbitrary δ, we have

$$\frac{d}{dt}\frac{\partial L}{\partial \dot{x}} = \frac{\partial L}{\partial x}. \tag{2.41}$$

As an example considering a spring mass system with mass m and spring constant k, then

$$L(x, \dot{x}) = \frac{1}{2}(m\dot{x}^2 - kx^2). \tag{2.42}$$

Then, Euler Lagrange equations read

$$m\ddot{x} = -kx. \tag{2.43}$$

2.3.2 Klein-Gordon Field

Consider a scalar field with Lagrangian density

$$L = \frac{1}{2}(\partial_\mu \phi \partial^\mu \phi - m\phi^2), \tag{2.44}$$

Given the action

$$S = \int L\, d^3x \tag{2.45}$$

Then

$$\delta S = \int \partial_\mu \phi \partial^\mu \delta\phi - m\phi\delta\phi \, d^3x \tag{2.46}$$

$$\delta S = \int (-\partial_\mu \partial^\mu \phi - m\phi)\delta\phi \, d^3x \tag{2.47}$$

where we integrate by parts with variation zero at boundary and we get for arbitrary $\delta\phi$, it should be true that

$$\partial_\mu \partial^\mu \phi + m^2 \phi = 0, \tag{2.48}$$

or

$$\frac{\partial^2 \phi}{\partial t^2} - \nabla^2 \phi + m^2 \phi = 0. \tag{2.49}$$

2.3.3 Variation of A Field

We equip A with a dynamics by defining Lagrangian as density

$$L = -\frac{\epsilon_0}{4} F_{\mu\nu} F^{\mu\nu}. \tag{2.50}$$

$$S = \int L \, d^3x \tag{2.51}$$

Then

$$\delta S = \int F_{\mu\nu}(\partial^\mu \delta A^\nu - \partial^\nu \delta A^\mu) \, d^3x, \tag{2.52}$$

$$\delta S = \int \partial^\mu F_{\mu\nu} \delta A^\nu \, d^3x, \tag{2.53}$$

where we used integration by parts and zero variation at the boundary to get

$$\partial^\mu F_{\mu\nu} = 0. \tag{2.54}$$

These are four beautiful Maxwell's equations. If we transform $A = \Lambda A'$, then we see that Maxwell's equations are satisfied in the transformed frame, that says it must indeed be the way that A transforms. This transformation gives $F^{\mu\nu} = \Lambda F'^{\mu\nu} \Lambda^T$.

Invoking the Lorentz gauge, we get

$$\partial^{\mu}\partial_{\mu}A_{\nu} = 0, \tag{2.55}$$

i.e.,

$$(\frac{\partial^2}{c^2\partial t^2} - \nabla^2)A_{\nu} = 0. \tag{2.56}$$

The solution is for $\varepsilon^{\mu} = (\varepsilon^0, \boldsymbol{\varepsilon})$, we have

$$A = \varepsilon \cos(\mathbf{k} \cdot \mathbf{r} - \omega t), \tag{2.57}$$

is a wave propagating in $\mathbf{k} = (k_x, k_y, k_z)$ direction with $\omega = c|\mathbf{k}|$. with $k^{\mu} = (\frac{\omega}{c}, \mathbf{k})$. The Lorentz gauge condition then becomes

$$k_{\mu}\varepsilon^{\mu} = 0. \tag{2.58}$$

For example,

$$A = \varepsilon \cos(k \cdot z - \omega t), \tag{2.59}$$

is a wave propagating in z direction with $\omega = c|k|$. From 2.36, the energy of this A field is $(\varepsilon_2^2 + \varepsilon_3^2)\frac{\epsilon_0\omega^2}{2c^2}V$. Therefore, for $\varepsilon_2^2 + \varepsilon_3^2 = 1$, we have

$$\begin{aligned} A &= c\sqrt{\frac{2\hbar}{V\epsilon_0\omega}}\varepsilon \cos(k \cdot z - \omega t) \\ &= c\sqrt{\frac{\hbar}{2\epsilon_0\omega V}}\varepsilon \text{ (exp } i(k \cdot z - \omega t) + \text{exp } -i(k \cdot z - \omega t) \text{)} \end{aligned} \tag{2.60}$$

has energy $\hbar\omega$, and this elementary excitation is termed **photon**. More generally, when ε is complex

$$A = c\sqrt{\frac{\hbar}{2\epsilon_0\omega V}} (\varepsilon \text{ exp } i(k \cdot z - \omega t) + \varepsilon^* \text{ exp } -i(k \cdot z - \omega t) \text{)} \tag{2.61}$$

For instance, if $\varepsilon = \frac{1}{\sqrt{2}}\begin{bmatrix}0\\1\\i\\0\end{bmatrix}$, we have

$$A = c\sqrt{\frac{\hbar}{\epsilon_0 \omega V}}\begin{bmatrix}0\\ \cos(k\cdot z - \omega t)\\ \sin(k\cdot z - \omega t)\\ 0\end{bmatrix} \tag{2.62}$$

constitutes circularly polarized light. If we move into a frame that rotates around z axis with angular velocity $\Delta\omega$, find the A transforms to

$$A' = c\sqrt{\frac{\hbar}{\epsilon_0 \omega V}}\begin{bmatrix}0\\ \cos(k\cdot z - (\omega+\Delta\omega)t)\\ \sin(k\cdot z - (\omega+\Delta\omega)t)\\ 0\end{bmatrix} \tag{2.63}$$

If we calculate the energy of A', we get two contributions, one due to z dependence of $\frac{\hbar\omega}{2}$ and t dependence, which is $\frac{\hbar(\omega+\Delta\omega)^2}{2\omega}$. Going from A to A', the energy changes by $\Delta E = \hbar\Delta\omega$, and the angular momentum is just $\frac{\Delta E}{\Delta\omega} = \hbar$. Thus, circularly polarized light carries angular momentum of $\boldsymbol{\hbar}$.

2.4 Electron

We now come to relativistic equation of an electron. Recall Pauli matrices

$$\sigma_x = \begin{bmatrix}0&1\\1&0\end{bmatrix};\ \sigma_y = \begin{bmatrix}0&-i\\i&0\end{bmatrix};\ \sigma_z = \begin{bmatrix}1&0\\0&-1\end{bmatrix};\ \mathbf{1} = \begin{bmatrix}1&0\\0&1\end{bmatrix} \tag{2.64}$$

For an electron wave $\exp(i(k\cdot r - \omega t)$, we want relativistic dispersion relation:

$$\hbar\omega = \sqrt{(m_o c^2)^2 + (c\hbar k_x)^2 + (c\hbar k_y)^2 + (c\hbar k_z)^2}.$$

For this, we make the wave a four vector $\psi = \exp(i(k\cdot r - \omega t)u$, where u is $+1$ eigenvector of $\cos\theta\alpha_\mu + \sin\theta\beta$, where $\sin\theta = \frac{m_0 c^2}{E}$ and $\beta = \sigma_x \otimes \mathbf{1}$ and $\alpha_u = -\sigma_z \otimes \sigma_\mu$.

ψ satisfies

$$i\hbar\frac{\partial\psi}{\partial t} = (mc^2\beta - ic\hbar\alpha_j\partial_j)\psi.$$

Using $\hbar = c = 1$ and multiplying both sides with β gives

$$(i\gamma^\mu \partial_\mu - m)\psi = 0,$$

where $\gamma^0 = \sigma_x \otimes \mathbf{1}$ and $\gamma^\mu = i\sigma_y \otimes \sigma_\mu$. Let us diagonalize the matrix:

$$C = m\beta + p_j\alpha_j = E(-\cos\theta\sigma_z \otimes \sigma_\alpha + \sin\theta\sigma_x \otimes \mathbf{1}) \tag{2.65}$$

$$= E\exp(i\frac{\theta}{2}\sigma_y \otimes \sigma_\alpha)(-\sigma_z \otimes \sigma_\alpha)\exp(-i\frac{\theta}{2}\sigma_y \otimes \sigma_\alpha) \tag{2.66}$$

Let $\xi_\alpha^\pm$ be eigenvectors of σ_α with eigenvalues ± 1. Then, $\begin{bmatrix} 1 \\ 0 \end{bmatrix} \otimes \xi_\alpha^\pm$ are eigenvectors of $-\sigma_z \otimes \sigma_\alpha$ with eigenvalues ± 1 and $\begin{bmatrix} 0 \\ 1 \end{bmatrix} \otimes \xi_\alpha^\pm$ are eigenvectors of with eigenvalues ∓ 1. Then,

$$\exp(i\frac{\theta}{2}\sigma_y \otimes \sigma_\alpha)\begin{bmatrix} 1 \\ 0 \end{bmatrix} \otimes \xi_\alpha^\pm = \begin{bmatrix} \cos\frac{\theta}{2} \\ \pm\sin\frac{\theta}{2} \end{bmatrix} \otimes \xi_\alpha^\pm \tag{2.67}$$

are eigenvectors of C with eigenvalues $\pm E$.

$$\exp(i\frac{\theta}{2}\sigma_y \otimes \sigma_\alpha)\begin{bmatrix} 0 \\ 1 \end{bmatrix} \otimes \xi_\alpha^\pm = \begin{bmatrix} \mp\sin\frac{\theta}{2} \\ \cos\frac{\theta}{2} \end{bmatrix} \otimes \xi_\alpha^\pm \tag{2.68}$$

are eigenvectors of C with eigenvalues $\mp E$. Let u and v be these eigenvectors with $\pm$ eigenvalues, respectively. Then, let

$$u_1(p) = \begin{bmatrix} \cos\frac{\theta}{2} \\ \sin\frac{\theta}{2} \end{bmatrix} \otimes \xi_\alpha^+; \qquad u_2(p) = \begin{bmatrix} \sin\frac{\theta}{2} \\ \cos\frac{\theta}{2} \end{bmatrix} \otimes \xi_\alpha^-; \tag{2.69}$$

$$v_1(p) = \begin{bmatrix} -\sin\frac{\theta}{2} \\ \cos\frac{\theta}{2} \end{bmatrix} \otimes \xi_\alpha^+; \qquad v_2(p) = \begin{bmatrix} \cos\frac{\theta}{2} \\ -\sin\frac{\theta}{2} \end{bmatrix} \otimes \xi_\alpha^- \tag{2.70}$$

where $1, 2$ represent positive and negative helicity, respectively.

2.4.1 Completeness Relation

Then, using

$$\sum_{s=1,2} u_s(p)u_s(p)^\dagger + v_s(p)v_s(p)^\dagger = \mathbf{1} \tag{2.71}$$

and

$$E_p \sum_{s=1,2} u_s(p)u_s(p)^\dagger - v_s(p)v_s(p)^\dagger = (mc^2\beta + c\,\alpha_i p_i) \tag{2.72}$$

This gives for $\mathbf{m} = mc^2$

$$\sum_{s=1,2} u_s(p)\bar{u}_s(p) = \frac{\not p + \mathbf{m}}{2E_p} \tag{2.73}$$

$$\sum_{s=1,2} v_s(p)\bar{v}_s(p) = \frac{\not p - \mathbf{m}}{2E_p}. \tag{2.74}$$

where $\not p = p_\mu \gamma^\mu$.

Coming back to Dirac equation

$$(\gamma^\mu p_\mu + m)u(p) = 0. \tag{2.75}$$

Let momentum p' is related to p by a Lorentz transformation. The Lorentz transformation that takes p to p' can be written as boost from p to rest and arbitrary rotation and then boost from rest to p'. We can represent this as

$$\Lambda = B_{p'}(\zeta')\exp(\theta_1\Omega_\alpha)B_p(-\zeta), \tag{2.76}$$

where B_x is boost along x direction. On spinor, it takes the form:

$$u(p') = \Sigma u(p), \tag{2.77}$$

where

$$\Sigma = \exp(-\frac{\zeta'}{2}\sigma_z \otimes \sigma_{p'})\exp(i\frac{\theta_1}{2}\mathbf{1} \otimes \sigma_\alpha)\exp(\frac{\zeta}{2}\sigma_z \otimes \sigma_p). \tag{2.78}$$

Then, it can be verified that

$$(\gamma^\mu p'_\mu + m)u(p') = 0. \tag{2.79}$$

It should be noted that Eq. 2.77 does not preserve norm. However, what is true is that if we normalize the spinors such that $\mathbf{u}(p) = \sqrt{E_p}u(p)$ then under Lorentz transformation

$$\Sigma\mathbf{u}(p) = \mathbf{u}(p'). \tag{2.80}$$

To see this, let

$$u(p) = \begin{bmatrix} \cos\frac{\theta(p)}{2} \\ \sin\frac{\theta(p)}{2} \end{bmatrix} \otimes \xi_p. \tag{2.81}$$

Consider a spinor $\mathbf{u}(0) = \sqrt{\frac{m}{2}}\begin{bmatrix} 1 \\ 1 \end{bmatrix} \times \xi_p$ at rest. If we boost it to momentum p, we can write $u(0) \rightarrow \mathbf{w}$:

$$\mathbf{w} = \exp(-\frac{\zeta}{2}\sigma_z \otimes \sigma_p)u(0) = \frac{1}{\sqrt{2}}\begin{bmatrix} \exp(-\frac{\zeta}{2}) \\ \exp(\frac{\zeta}{2}) \end{bmatrix} \times \xi_p. \tag{2.82}$$

Since $\mathbf{w} \propto u(p)$, we have

$$\tan\frac{\theta(p)}{2} = \exp(\zeta). \tag{2.83}$$

Then

$$|\mathbf{w}|^2 = m\cosh(\zeta) = \frac{m}{\sin\theta(p)} = E_p, \tag{2.84}$$

but this says that

$$\mathbf{u}(p) = \mathbf{w}. \tag{2.85}$$

Then, from 2.78

$$\Sigma\,\mathbf{u}(p) = \exp(-\frac{\zeta'}{2}\sigma_z \otimes \sigma_{p'})\exp(i\frac{\theta_1}{2}\mathbf{1} \otimes \sigma_\alpha)\,\mathbf{u}(0), \tag{2.86}$$

$$= \sqrt{m}\,\exp(-\frac{\zeta'}{2}\sigma_z \otimes \sigma_{p'})(a\frac{1}{\sqrt{2}}\begin{bmatrix} 1 \\ 1 \end{bmatrix} \otimes \xi^+_{p'} + b\frac{1}{\sqrt{2}}\begin{bmatrix} 1 \\ 1 \end{bmatrix} \otimes \xi^-_{p'}), \tag{2.87}$$

where $a^2 + b^2 = 1$. Then

$$\Sigma\,\mathbf{u}(p) = \sqrt{m}\,(a\frac{1}{\sqrt{2}}\begin{bmatrix} \exp(-\frac{\zeta'}{2}) \\ \exp(\frac{\zeta'}{2}) \end{bmatrix} \otimes \xi^+_{p'} + b\frac{1}{\sqrt{2}}\begin{bmatrix} \exp(\frac{\zeta'}{2}) \\ \exp(\frac{-\zeta'}{2}) \end{bmatrix} \otimes \xi^-_{p'}), \tag{2.88}$$

$$\propto (a\begin{bmatrix} \cos\frac{\theta(p')}{2} \\ \sin\frac{\theta(p')}{2} \end{bmatrix} \otimes \xi^+_{p'} + b\begin{bmatrix} \sin\frac{\theta(p')}{2} \\ \cos\frac{\theta(p')}{2} \end{bmatrix} \otimes \xi^-_{p'}). \tag{2.89}$$

Then, equating coefficients of $\xi_{p'}$, we get $\propto$ in last equation is $\sqrt{E_{p'}}$ and

$$\Sigma \mathbf{u}(p) = \mathbf{u}(p'). \tag{2.90}$$

2.5 Electron-Photon Interaction

2.5.1 Electric-Magnetic Field Lagrangian and Hamiltonian

A charged particle with mass m and charge q in electric-magnetic field has Lagrangian

$$L = \frac{m}{2}(\dot{x}^2 + \dot{y}^2 + \dot{z}^2) + q(\dot{x}A_x + \dot{y}A_y + \dot{z}A_z) - qA_0, \tag{2.91}$$

where $\mathbf{A}$ is vector potential and $B = \nabla \times \mathbf{A}$, i.e.,

$$B_x = \frac{\partial A_z}{\partial y} - \frac{\partial A_y}{\partial z}, \tag{2.92}$$

$$B_y = \frac{\partial A_x}{\partial z} - \frac{\partial A_z}{\partial x}, \tag{2.93}$$

$$B_z = \frac{\partial A_y}{\partial x} - \frac{\partial A_x}{\partial y}. \tag{2.94}$$

$$E_x = -\frac{\partial A_x}{\partial t} - \frac{\partial A_0}{\partial x}, \tag{2.95}$$

$$E_y = -\frac{\partial A_y}{\partial t} - \frac{\partial A_0}{\partial y}, \tag{2.96}$$

$$E_z = -\frac{\partial A_z}{\partial t} - \frac{\partial A_0}{\partial z}. \tag{2.97}$$

We have Euler Lagrange equations:

$$m\ddot{x} + q\dot{A_x} = q(\dot{x}\frac{\partial A_x}{\partial x} + \dot{y}\frac{\partial A_y}{\partial x} + \dot{z}\frac{\partial A_z}{\partial x}) - q\frac{\partial A_0}{\partial x}. \tag{2.98}$$

Writing

$$\dot{A_x} = \frac{\partial A_x}{\partial t} + \dot{x}\frac{\partial A_x}{\partial x} + \dot{y}\frac{\partial A_x}{\partial y} + \dot{y}\frac{\partial A_x}{\partial y}. \tag{2.99}$$

Substituting in 2.98, we get

$$m\ddot{x} = q(\dot{y}B_z - \dot{z}B_y) - q(\frac{\partial A_x}{\partial t} + \frac{\partial A_0}{\partial x}), \tag{2.100}$$

and similarly for y, z gives in all that

$$m\dot{v} = q(v \times B + E). \tag{2.101}$$

The *Lorentz* force law. The momentum $p_x = \frac{\partial L}{\partial \dot{x}} = m\dot{x} + qA_x$ and similarly for y, z gives the Hamiltonian or energy:

$$\begin{aligned} H &= \dot{x}p_x + \dot{y}p_y + \dot{z}p_z - L \\ &= \frac{(p_x - qA_x)^2}{2m} + \frac{(p_y - qA_y)^2}{2m} + \frac{(p_z - qA_z)^2}{2m} + qA_0. \end{aligned} \tag{2.102}$$

The energy E is then

$$E - qA_0 = \frac{(p_x - qA_x)^2}{2m} + \frac{(p_y - qA_y)^2}{2m} + \frac{(p_z - qA_z)^2}{2m}. \tag{2.103}$$

The relativistic generalization is

$$E - qA_0 = c\sqrt{(p_x - qA_x)^2 + (p_y - qA_y)^2 + (p_z - qA_z)^2 + (mc)^2}. \tag{2.104}$$

We use this energy to define Dirac equation in the electric-magnetic field.

2.5.2 *Gauge Coupling and Transitions*

Recall how the Dirac equation reads with $p^\mu = (E, \mathbf{p}c)$, we have

$$(\gamma^\mu p_\mu - m)\psi = 0 \tag{2.105}$$

From 2.104, we have the Dirac equation in the presence of electromagnetic field as $p_\mu \to p_\mu - qA_\mu$. If we identify p_μ with $i\partial_\mu$, the Dirac equation reads

$$(i\gamma^\mu \partial_\mu - mc - qA_\mu \gamma^\mu)\psi = 0 \tag{2.106}$$

or in terms of matrices α_j, β

$$i\partial_t \psi = (-c\alpha_j\, i\partial_j + mc^2\beta - qA_j\alpha_j + qA_0)\psi \tag{2.107}$$

without A, we have the free Dirac equation

$$i\partial_t \psi = (-c\alpha_j\, i\partial_j + mc^2\beta)\psi \tag{2.108}$$

The term

$$T = -qA_j\alpha_j + qA_0, \tag{2.109}$$

is **transition term**. In absence of this if electron is in eigenstate of the Dirac equation $\psi_1 = u_1(p)\exp(ip\cdot r)$, it will stay in this state. In the presence of the transition term, it makes a transition. Let's imagine a photon is present, and then A is as in Eq. 2.110:

$$\begin{aligned} A &= c\sqrt{\frac{2\hbar}{\epsilon_0\omega V}}\varepsilon\cos(k\cdot r-\omega t) \\ &= c\sqrt{\frac{\hbar}{2\epsilon_0\omega}}\varepsilon\,(\,\exp\ i(k\cdot r-\omega t)+\exp\ -i(k\cdot r-\omega t)\,) \end{aligned} \tag{2.110}$$

Then T acting on ψ_1 induces change $u_1(p)\exp(ip\cdot r) \to u_1(p+k)\exp(i(p+k)\cdot r)$. Let's call the state $u_1(p+k)\exp(i(p+k)\cdot r)$ as ψ_2. Let x_1 and x_2 denote coefficients of ψ_1 and ψ_2. Then

$$\frac{d}{dt}\begin{bmatrix} x_1 \\ x_2 \end{bmatrix} = \frac{-i}{\hbar}\begin{bmatrix} E_p & \Omega^\dagger\exp(i\omega t) \\ \Omega\exp(-i\omega t) & E_{p+k} \end{bmatrix}\begin{bmatrix} x_1 \\ x_2 \end{bmatrix} \tag{2.111}$$

where E_P and E_{p+k} are energies of the initial and final electron states and

$$\begin{aligned} \Omega &= qc\sqrt{\frac{\hbar}{2V\epsilon_0\omega}}\,u_1^\dagger(p+k)(-\epsilon_j\alpha_j+\epsilon_0)u_1(p) \\ &= \frac{C}{\sqrt{2E_k}}\bar{u}_1(p+k)(\epsilon_\mu\gamma^\mu)u_1(p), \end{aligned} \tag{2.112}$$

where $\bar{u} = u^\dagger\beta$, $E_k = \hbar\omega$ the energy of the photon, and

$$C = \frac{qc\hbar}{\sqrt{V\epsilon_0}}. \tag{2.113}$$

If we denote $\tilde{x}_1 = \exp(-i\omega t)x_1$, then

$$\frac{d}{dt}\begin{bmatrix} \tilde{x}_1 \\ x_2 \end{bmatrix} = \frac{-i}{\hbar}\begin{bmatrix} E_p+E_k & \Omega^* \\ \Omega & E_{p+k} \end{bmatrix}\begin{bmatrix} \tilde{x}_1 \\ x_2 \end{bmatrix} \tag{2.114}$$

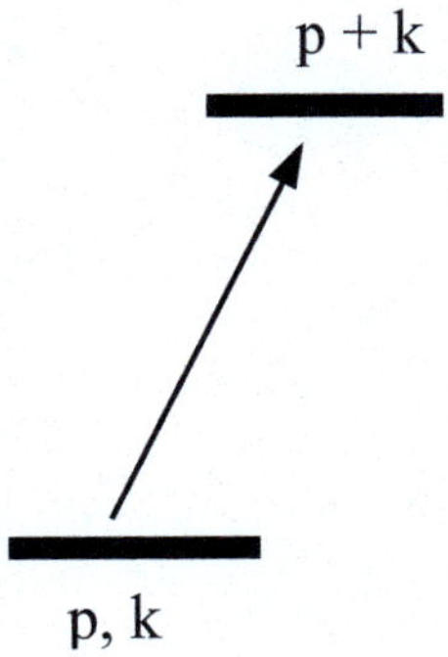

Fig. 2.2 Figure shows transitions between electron, photon states

where $E_k = \hbar\omega_k$ the energy of the photon. Observe $\tilde{x}_1$ is coefficient of state evolving with energy $E_p + E_k$, the joint state of electron and photon. d then represents the transition out of this state. The initial state has electron and photon, while the final state only has electron. We denote this as

$$|p, k\rangle \rightarrow |p + k\rangle \tag{2.115}$$

Drawn as an energy level diagram, it looks like Fig. 2.2.

We showed transition to state $u_1(p + k)$; similarly, we have transition to state $u_2(p + k)$, etc., where 1, 2 represent helicity.

2.5.3 Feynman Diagrams

The above level diagram is also represented by so-called Feynman diagram as shown in Fig. 2.3. An electron and photon of momentum p and k, respectively, react to form an electron of momentum $p + k$.

This reaction is very unfavorable for large energy difference between input and out states of the two levels as shown in Fig. 2.2. However, the outgoing electron can immediately dissociate into an electron and photon of momentum p' and k such that $p' + k' = p + k$. Furthermore, $E_P + E_k = E_{p'} + E_{k'}$, so that the initial and final states have same energy momentum. Then, the overall reaction which is represented by a level diagram in Fig. 2.4a and Feynman diagram 2.4b becomes favorable. We can compute the rate of this reaction, which we do in the next chapter on quantum electrodynamics a subject that calculates such reaction rates.

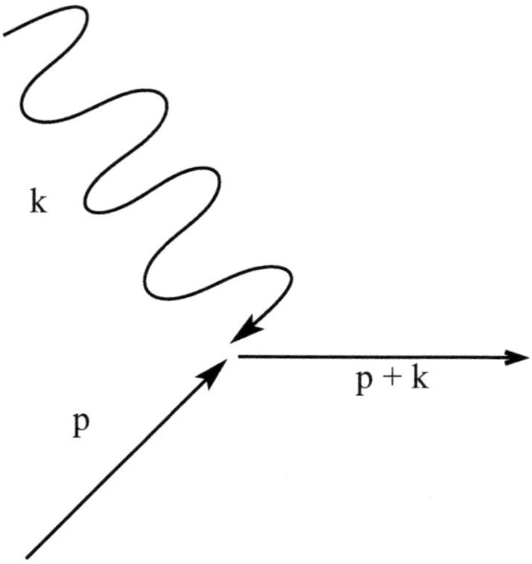

Fig. 2.3 Figure shows a Feynman diagram of electron and photon of momentum p and k respectively react to form an electron of momentum $p + k$

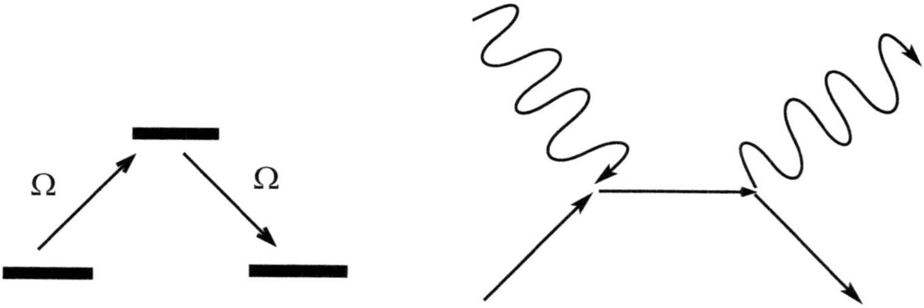

Fig. 2.4 (**a**) shows a three-level diagram and (**b**) shows a Feynman diagram for electron-photon scattering

2.6 Lorentz Gauge vs. $E \cdot x$ Gauge

For a plane wave along z direction, with electric field $E_x \sin(kz - \omega t)$, the Lorentz gauge is

$$(A_0, A_x, A_y, A_z) = \frac{E_x}{\omega} \cos(kz - \omega t)(0, 1, 0, 0).$$

But this gauge is not suited for calculating optical transitions, because we don't recover the Rabi frequency $qE_x d$ (d electric dipole moment). What we find is something orders of magnitude smaller. Nor is it suitable for calculating electron-electron scattering as described in the next chapter on quantum electrodynamics (QED), because we don't recover Coulomb potential. What we find is something orders of magnitude smaller. Instead, we work with $E \cdot x$ gauge:

$$(A_0, A_x, A_y, A_z) = \frac{E_x}{2}(-x \ \sin(kz - \omega t), \frac{\cos(kz - \omega t)}{\omega}, 0, -\frac{x}{c} \sin(kz - \omega t)) \tag{2.116}$$

(c light velocity) to find everything correct. What we get is a new propagator that describes amplitude of electron-electron scattering which gives us Coulomb potential.

A more general gauge is

$$\begin{aligned}(A_0, A_x, A_y, A_z) = E_x(&-\cos^2\theta x \ \sin(kz - \omega t), \sin^2\theta \tfrac{\cos(kz-\omega t)}{\omega}, 0, \\ &-\cos^2\theta \tfrac{x}{c} \sin(kz - \omega t)),\end{aligned} \tag{2.117}$$

we will find that we use this general gauge for successful renormalization in the next chapter, when photon wavelength is very small compared to Compton wavelength.

Let's see why in optical transitions, Rabi frequency goes bad, if we do not use right gauge. This equation is not very tractable, because it is nonlinear in A; let's write a linear equation, which is the Dirac equation, which takes the form

$$i\frac{\partial \phi}{\partial t} = \left(\sum_{j=x,y,z} c(-i\,\hbar\frac{\partial}{\partial x_j} - qA_j)\alpha_j + \beta mc^2 + qA_0 \right) \phi. \tag{2.118}$$

where $\alpha_j = \sigma_z \otimes \sigma_j$ and $\beta = \sigma_x \otimes \mathbf{1}$ are Dirac matrices, where σ_j are the Pauli matrices, $\sigma_z = \begin{pmatrix} 1 & 0 \\ 0 & -1 \end{pmatrix}$. ϕ is electron spinor, for a electron wave with momentum k, takes the form $\phi = \begin{bmatrix} \cos\frac{\theta}{2} \\ \sin\frac{\theta}{2} \end{bmatrix} \otimes \uparrow$, where $\uparrow$ is spin up, $\cos\theta = \frac{\hbar k}{mc} = \frac{\upsilon}{c}$, where $\upsilon = \frac{\hbar k}{m}$, is electron wave group velocity. Electron orbitals are of size $\sim A^\circ$, their $k \sim 10^{10}m$, and then $\upsilon \sim 10^6 m/s$ and $\cos\theta \sim \frac{10^6}{3\times 10^8} \sim 10^{-3}$. Electron is nonrelativistic, $\cos\theta = \frac{\upsilon}{c} \sim 0, \theta \sim \frac{\pi}{2}, \phi = \frac{1}{\sqrt{2}}\begin{bmatrix} 1 \\ 1 \end{bmatrix} \otimes \uparrow$.

To fix ideas, take incoming EM wave, along z direction, with electric field $E_x \sin(kz - \omega t)$, the Lorentz gauge is $(A_0, A_x, A_y, A_z) = \frac{E_x}{\omega}\cos(kz - \omega t)(0, 1, 0, 0)$. Electron wave with momentum q absorbs the photon with momentum k and transits to momentum $q + k$. The transition is driven by Dirac matrix α_x, with transition amplitude:

$$\mathcal{M} = \begin{bmatrix} \cos\frac{\theta}{2} \\ \sin\frac{\theta}{2} \end{bmatrix} \otimes \uparrow \underbrace{(\sigma_z \otimes \sigma_x)}_{\alpha_x} \begin{bmatrix} \cos\frac{\theta}{2} \\ \sin\frac{\theta}{2} \end{bmatrix} \otimes \downarrow = qcA_x\frac{\upsilon}{c} = qE_x\frac{\upsilon}{\omega} \tag{2.119}$$

If we have electron orbital ϕ_0, then $k' = \frac{M}{M+m}k$ of photon momentum goes to electron-nuclear relative coordinate, while $k'' = k$ momentum to CM (center of mass), where M is nucleus mass. The process drives the transition:

$$\phi_0 \uparrow \longrightarrow \exp(ik'z)\phi_0 \downarrow,$$

with amplitude $\mathcal{M} = qE_x\frac{v}{\omega}$.

When orbital ϕ_1 is different from ϕ_0, we go to

$$\phi_0 \uparrow \longrightarrow \exp(ik'z)\phi_0 \downarrow,$$

with amplitude $\mathcal{M} = qcA_x\frac{v}{c}$ whose overlap with ϕ_1 is

$$\mathcal{M}_1 = qcA_x\frac{v}{c}ik' \underbrace{\langle\phi_1|z|\phi_0\rangle}_{d_z} = iqE_xd_z\,\frac{v}{c},$$

where $ck' \sim \omega$.

But this is not suited for study of optical transitions, because we don't recover the Rabi frequency qE_xd. What we find is orders of magnitude smaller (down by $\frac{v}{c}$). Instead, we work with gauge

$$(A_0, A_x, A_y, A_z) = \frac{-E_x}{2}(x\,\sin(kz-\omega t), -\frac{\cos(kz-\omega t)}{\omega}, 0, \frac{x}{c}\sin(kz-\omega t)).$$

Now, we have process driven by x term. For the e process, the amplitude of $\phi_0 \to \phi_0$ is just 0, as $\langle\phi_0|x|\phi_0\rangle = 0$ and the amplitude of $\phi_0 \to \phi_1$ is simply

$$\mathcal{M}_1' = qE_x \underbrace{\langle\phi_1|x|\phi_0\rangle}_{d_x} = qE_xd_x,$$

Dipole elements d_z, d_x are approx. Bohr radius $\sim A^\circ$. Due to the factor $\frac{v}{c} \sim 10^{-3}$, $\mathcal{M}_1 \ll \mathcal{M}_1'$. Therefore, transition between different atomic orbitals is largely driven by the x term.

2.7 Problems

1. Frame O' moves with respect to frame O with velocity u along x axis. If velocity of a particle in frame O' is $v' = (v_x', v_y', v_z')$, find its velocity in frame O.
2. The rest mass of a proton is 938 Mev/c^2. If it moves with velocity 0.9 c, where c is velocity of light, find its energy.
3. In above problem if the proton moves such that its momentum is 1000 Mev/c, find its energy.

4. Muon at rest, decay in time 2.2 μsec. How much time will it take them to decay if they are moving at velocity 0.99 c.
5. Lagrangian of electromagnetic field is

$$L = \frac{-\epsilon_0}{4} \int F_{\mu\nu} F^{\mu\nu} d^3x$$

By taking variation of L, show that we get Maxwell's equations $\partial^{\mu} F_{\mu\nu} = 0$.

Chapter 3
Quantum Electrodynamics

3.1 Introduction

We first develop an analogy between the three-level atomic system, the so-called Λ system and scattering processes in quantum electrodynamics (QED) [1–4]. In a Λ system as shown in Fig. 3.1, we have two ground state levels $|1\rangle$ and $|3\rangle$ at energy E_1 and excited level $|2\rangle$ at energy E_2. The transition from $|1\rangle$ to $|2\rangle$ has strength Ω_1 and transition from $|2\rangle$ to $|3\rangle$ has strength Ω_2. In the interaction frame of natural Hamiltonian system, we get a second-order term connecting level $|1\rangle$ to $|3\rangle$ with strength $\frac{\Omega_1\Omega_2}{(E_1-E_2)}$. This term creates an effective coupling between ground state levels and drives transition from $|1\rangle$ to $|3\rangle$. Scattering processes in QED can be modelled like this. Feynman amplitudes are calculation of second-order term $\mathcal{M} = \frac{\Omega_1\Omega_2}{(E_1-E_2)}$.

The state of the three-level system evolves according to the Schrödinger equation

$$\dot{\psi} = \frac{-i}{\hbar}\begin{bmatrix} E_1 & \Omega_1^* & 0 \\ \Omega_1 & E_2 & \Omega_2^* \\ 0 & \Omega_2 & E_1 \end{bmatrix}\psi. \tag{3.1}$$

We proceed into the interaction frame of the natural Hamiltonian (system energies) by transformation:

$$\phi = \exp(\frac{i}{\hbar}\begin{bmatrix} E_1 & 0 & 0 \\ 0 & E_2 & 0 \\ 0 & 0 & E_1 \end{bmatrix})\psi. \tag{3.2}$$

N. Khaneja, *High Energy Physics*, https://doi.org/10.1007/978-3-031-75645-0_3

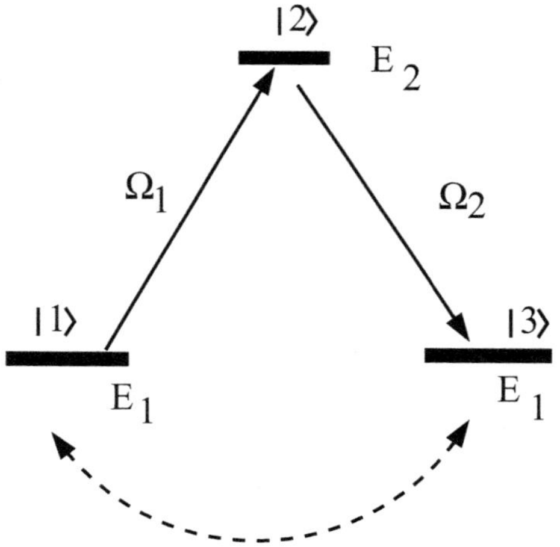

Fig. 3.1 Figure shows a three-level Λ system, with two ground state levels $|1\rangle$ and $|3\rangle$ and an excited level $|2\rangle$

This gives for $\Delta E = E_2 - E_1$

$$\dot{\phi} = \underbrace{\frac{-i}{\hbar}\begin{bmatrix} 0 & \exp(-\frac{i}{\hbar}\Delta E\ t)\Omega_1^* & 0 \\ \exp(\frac{i}{\hbar}\Delta E\ t)\Omega_1 & 0 & \exp(\frac{i}{\hbar}\Delta E\ t)\Omega_2^* \\ 0 & \exp(-\frac{i}{\hbar}\Delta E\ t)\Omega_2 & 0 \end{bmatrix}}_{H(t)}\phi. \tag{3.3}$$

$H(t)$ is periodic with period $\Delta t = \frac{2\pi}{\Delta E}$. After Δt, the system evolution is

$$\phi(\Delta t) = (I + \int_0^{\Delta t} H(\sigma)d\sigma + \int_0^{\Delta t}\int_0^{\sigma_1} H(\sigma_1)H(\sigma_2)d\sigma_2 d\sigma_1 + \ldots)\phi(0). \tag{3.4}$$

The first integral averages to zero, while the second integral

$$\int_0^{\Delta t}\int_0^{\sigma_1} H(\sigma_1)H(\sigma_2)d\sigma_2 d\sigma_1 = \frac{1}{2}\int_0^{\Delta t}\int_0^{\sigma_1} [H(\sigma_1), H(\sigma_2)]d\sigma_2 d\sigma_1. \tag{3.5}$$

Evaluating it explicitly, we get for our system that second-order integral is

$$\frac{-i\Delta t}{\hbar}\begin{bmatrix} 0 & & \frac{\Omega_1^*\Omega_2^*}{E_1 - E_2} \\ 0 & 0 & 0 \\ \underbrace{\frac{\Omega_1\Omega_2}{E_1 - E_2}}_{M} & 0 & 0 \end{bmatrix}. \tag{3.6}$$

Thus, we have created an effective Hamiltonian:

$$\begin{bmatrix} 0 & & \frac{\Omega_1^* \Omega_2^*}{E_1 - E_2} \\ 0 & 0 & 0 \\ \frac{\Omega_1 \Omega_2}{E_1 - E_2} & 0 & 0 \end{bmatrix}, \tag{3.7}$$

which couples level $|1\rangle$ and $|3\rangle$ and drives transition between them at rate $\mathcal{M} = \frac{\Omega_1 \Omega_2}{(E_1 - E_2)}$.

3.2 Coulomb Potential and Møller Scattering

The heart of interactions in high energy physics is the beautiful electron-electron scattering of Møller. The coulomb interaction between electrons. Figure 3.2 shows two electrons with momentum p and $-p$ scatter by exchange of photon say in z direction to $p + q$ and $-(p + q)$. The scattering amplitude is well known, given as Feynman propagator $\mathcal{M} = \frac{(e\hbar c)^2}{\epsilon_0 V} \frac{\bar{u}(p+q)\gamma^\mu u(p)\, \bar{u}(-(p+q))\gamma_\mu u(-p)}{q^2}$, [2, 3, 5, 7], where V is the volume of the scattering electrons, e elementary charge, and ϵ_0 permittivity of vacuum. But this needs to be taken with grain of salt. Since we exchange photon momentum in z direction, we have two photon polarization x, y, and hence, the true scattering amplitude should be $\mathcal{M}_1 =$

$$\frac{(e\hbar c)^2}{\epsilon_0 V} \frac{\begin{array}{c} \bar{u}(p+q)\gamma^x u(p)\, \bar{u}(-(p+q))\gamma_x u(-p) \\ +\bar{u}(p+q)\gamma^y u(p)\, \bar{u}(-(p+q))\gamma_y u(-p) \end{array}}{q^2}.$$

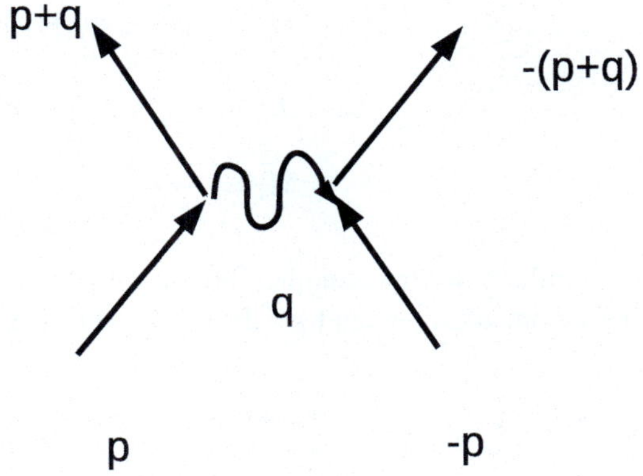

Fig. 3.2 Figure depicts Møller scattering. Two electrons with momentum p and $-p$ scatter by exchange of photon to $p + q$ and $-(p + q)$

But when electrons are nonrelativistic, $\mathcal{M}_1 \sim 0$. This is disturbing; how will we ever get the coulomb potential, where $\mathcal{M} \sim \frac{(e\hbar c)^2}{\epsilon_0 V q^2}$? Where is the problem? The problem is with the EM gauge used in Dirac equation.

For a plane wave along z direction, with electric field $E_x \sin(kz-\omega t)$, the Lorentz gauge is $(A_0, A_x, A_y, A_z) = \frac{E_x}{\omega}\cos(kz-\omega t)(0, 1, 0, 0)$. But this gauge is not suited for calculating optical transitions, because we don't recover the Rabi frequency $qE_x d$ (d electric dipole moment). What we find is something orders of magnitude smaller. Nor is it suitable for calculating electron-electron scattering because we don't recover coulomb potential. What we find is something orders of magnitude smaller. Instead, we work with $E \cdot x$ gauge

$$(A_0, A_x, A_y, A_z) = \frac{-E_x}{2}(x \sin(kz - \omega t), -\frac{\cos(kz - \omega t)}{\omega}, 0, \frac{x}{c}\sin(kz - \omega t))$$

(c light velocity) to find everything correct. What we get is a new Feynman propagator. Let's build up to it.

In Møller scattering, electrons with momentum p_1 and q_1 exchange photon with momentum k and scatter to new momentum states p_2 and q_2. Observe the virtual particle four momentum is k. The Feynman diagram for the process is in Fig. 3.3. There are two three-level systems associated with this process. Let $P = p_1 + q_1$.

In Fig. 3.3a, we have the first three-level system where the electron with momentum p_1 is annihilated, an electron of momentum p_2 is created, and a photon of momentum $q = p_1 - p_2$ is created. Subsequently, the electron with momentum q_1 is annihilated , an electron of momentum q_2 is created, and photon of momentum k is annihilated. The amplitude for this process is

$$\Omega_1 = \frac{C}{\sqrt{2E_q}}\,\bar{u}(p_2)\gamma^{\nu}\epsilon_{\nu}^{*}(q)u(p_1), \tag{3.8}$$

$$\Omega_2 = \frac{C}{\sqrt{2E_q}}\,\bar{u}(q_2)\gamma^{\nu}\epsilon_{\nu}(q)u(q_1), \tag{3.9}$$

$$E_1 - E_2 = E_{p_1} - E_{p_2} - E_q = q_0 - E_q, \tag{3.10}$$

$$\mathcal{M}_1 = \frac{\Omega_1\Omega_2}{(E_1 - E_2)}. \tag{3.11}$$

Similarly, we have another three-level system, Fig. 3.3b in which q_1 emits photon with momentum $-k$ and p_1 absorbs it. This gives

$$\Omega_1 = \frac{C}{\sqrt{2E_q}}\,\bar{u}(q_2)\gamma^{\nu}\epsilon_{\nu}(q)u(q_1), \tag{3.12}$$

$$\Omega_2 = \frac{C}{\sqrt{2E_q}}\,\bar{u}(p_2)\gamma^{\nu}\epsilon_{\nu}^{*}(q)u(p_1), \tag{3.13}$$

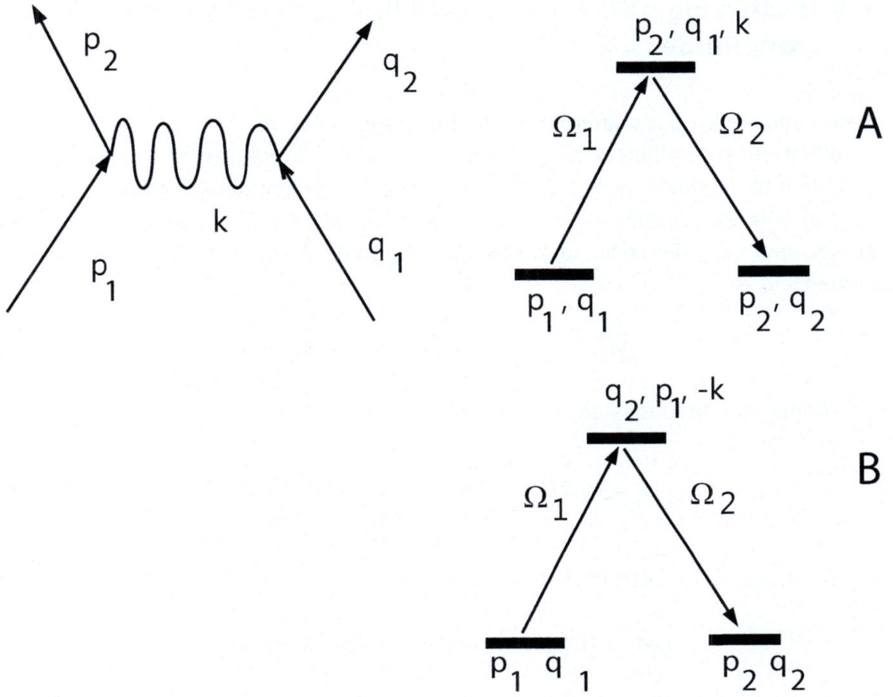

Fig. 3.3 Figure shows the Feynman diagram for the Møller scattering and its corresponding three-level system. The electron with momentum p_1 emits (absorbs) a photon and scatters to momentum p_2; the photon is absorbed (emitted) by electron with momentum q_1 which scatters to momentum q_2

$$E_1 - E_2 = E_{q_1} - E_{q_2} - E_q = -(q_0 + E_q), \tag{3.14}$$

$$\mathcal{M}_2 = \frac{\Omega_1 \Omega_2}{(E_1 - E_2)}, \tag{3.15}$$

where we used conservation of energy $E_{p_2} - E_{p_1} = E_{q_1} - E_{q_2}$. When we add the two amplitudes, we get for $q = p_1 - p_2$ and $q^2 = q_\mu q^\mu$, the total amplitude is

$$\frac{\Omega_1 \Omega_2}{q^2} = C^2 \frac{\bar{u}(q_2)\gamma^\nu \epsilon_\nu(q) u(q_1)\ \bar{u}(p_2)\gamma^\nu \epsilon_\nu^*(q) u(p_1)}{q^2} \tag{3.16}$$

We can now sum over photon polarization ϵ say along x and y axis to get

$$\begin{aligned} \mathcal{M} &= \mathcal{M}_x + \mathcal{M}_y \\ &= \frac{(\hbar c)^2}{\epsilon_0 V} \frac{\bar{u}(q_2)\gamma^x u(q_1)\ \bar{u}(p_2)\gamma^x u(p_1) + \bar{u}(q_2)\gamma^y u(q_1)\bar{u}(p_2)\gamma^y u(p_1)}{q^2} \end{aligned} \tag{3.17}$$

3.3 Scattering with $E \cdot x$ Term: The Negative Sign of Amplitude

Now, consider Moller scattering with x term (Fig. 3.4).

When electron changes momentum by q (say z direction), in Lorentz gauge, photon of momentum $-q$ is emitted. In e $\cdot\, x$ gauge, the emitted photon can be more general with momentum $-q + k$, where $k = n\Delta$ ($\Delta\, \text{ł} = 2\pi$, l length of electron packet) in x or y direction, and then, the amplitude $\mathcal{M}$ of scattering a momentum exchange q is

$$\mathcal{M}_0 = C u_4^\dagger(p_4) u_3(p_2)\, u_2^\dagger(p_3) u_1(p_1) \tag{3.18}$$

Adding two directions gives

$$C = 2\frac{(e\hbar c)^2}{\epsilon V}\frac{1}{4}\sum_n \frac{1}{n^2}\frac{1}{|q|^2} \sim \frac{(e\hbar c)^2}{\epsilon V |q|^2}, \tag{3.19}$$

We of course have (from A_z) the term:

$$\mathcal{M}_z = C u_4^\dagger(p_4)\gamma_z u_3(p_2)\, u_2^\dagger(p_3)\gamma^z u_1(p_1) \tag{3.20}$$

The total amplitude including contribution from A_x, A_y in gauge

$$\mathcal{M} = \frac{1}{4|q|^2}\{u_4^\dagger(p_4)\gamma_\mu u_3(p_2)\, u_2^\dagger(p_3)\gamma^\mu u_1(p_1) - (3\mathcal{M}_0 + 5\mathcal{M}_z)\}. \tag{3.21}$$

The first term is the usual Feynman propagator, scaled by $\frac{1}{4}$, but the second term is new and gives big contribution to coulomb potential.

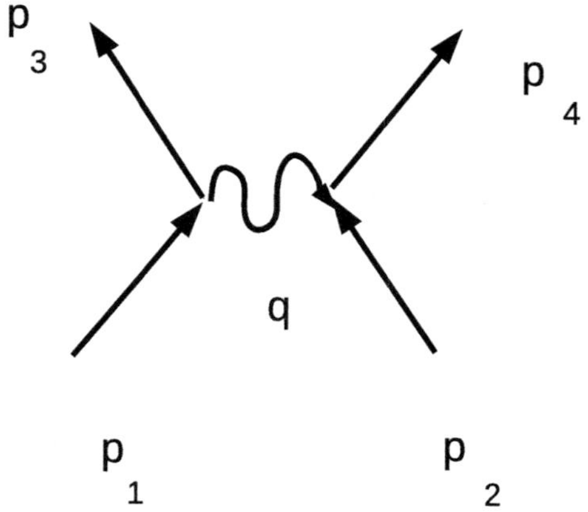

Fig. 3.4 Figure depicts Møller scattering. Two electrons with momentum p_1 and p_2 scatter by exchange of photon to p_3 and p_4

That's it; we have a new propagator. In its full glory, it reads

$$\mathcal{M} = \frac{(e\hbar c)^2}{4\epsilon_0 V q^2}\{u_4^\dagger(p_4)\gamma_\mu u_3(p_2)\, u_2^\dagger(p_3)\gamma^\mu u_1(p_1) - \left(3\mathcal{M}_0 + 5\mathcal{M}_q\right)\}. \tag{3.22}$$

What is remarkable we have been able to get $-\mathcal{M}_0$, which is difficult to explain in Feynman propagator. However, the propagator is not Lorentz invariant; we say our electron-phonon coupling is scaled such that the true propagator is

$$\mathcal{M} = \frac{(e\hbar c)^2}{(\frac{E_1+E_2}{2})^2\epsilon_0 V q^2}\{\mathbf{u}_4^\dagger(p_4)\gamma_\mu \mathbf{u}_3(p_2)\, \mathbf{u}_2^\dagger(p_3)\gamma^\mu \mathbf{u}_1(p_1)\}. \tag{3.23}$$

Now, this propagator is relativistically invariant; when we boost from CM frame to other frame (OF) , then $\mathcal{M}_{CM} = \frac{\mathcal{M}_{OF}}{\gamma}$. And defining the relativistic discount factor $\eta = \frac{\sqrt{E_1E_2E_3E_4}}{(\frac{E_1+E_2}{2})^2}$, we can write the propagator as $\mathcal{M} = \eta\mathcal{M}_F$ where $\mathcal{M}_F$ is Feynman propagator:

$$\mathcal{M}_F = \frac{(e\hbar c)^2}{\epsilon_0 V q^2}\{u_4^\dagger(p_4)\gamma_\mu u_3(p_2)\, u_2^\dagger(p_3)\gamma^\mu u_1(p_1)\}. \tag{3.24}$$

It is this discount factor that plays a important role in our treatment of QED. We will always calculate $\mathcal{M}_F$ and then apply the discount factor η. In CM, $\eta = 1$ amd $\mathcal{M}_{CM} = \mathcal{M}_F$. In CM

$$\mathcal{M}_q = \frac{e^2}{\epsilon_0 V}\frac{1}{|q|^2} \tag{3.25}$$

This gives a scattering potential:

$$V = \sum_q \mathcal{M}_q \exp(-iq(r_1 - r_2)) = \frac{e^2}{(2\pi)^3\epsilon_0}\int d^3q \frac{\exp(-ik(r_1 - r_2))}{|q|^2}. \tag{3.26}$$

For $r = r_1 - r_2$, we have

$$\begin{aligned}\int d^3q \frac{\exp(-iq(r_1 - r_2))}{|q|^2} &= 2\pi \int d|q| \int_0^\pi \exp(-i|q||r|\cos\theta)\sin\theta d\theta \\ &= \frac{4\pi}{|r|}\int_0^\infty \frac{\sin|q||r|}{|q|}d|q| = \frac{2\pi^2}{|r|}\end{aligned}$$

$$V = \sum_k \mathcal{M}_q \exp(-iq(r_1 - r_2)) = \frac{e^2}{4\pi\epsilon_0|r|} \tag{3.27}$$

The familiar Coulomb potential.

3.4 Bhabha Scattering

In quantum electrodynamics, Bhabha scattering is the electron-positron scattering process:

$$e^+e^- \rightarrow e^+e^- \tag{3.28}$$

Bhabha scattering is named after the Indian physicist Homi J. Bhabha. The Bhabha scattering rate is used as a luminosity monitor in electron-positron colliders (Fig. 3.5).

How do we understand scattering of an electron and positron.

3.4.1 *Annihilation*

See Fig. 3.6. A photon of momentum $p+k$ comes and strikes filled sea of negative energy electrons and ejects a negative energy electron of state k and momentum $-k$ and creates an electron of momentum p and leaves behind missing momentum

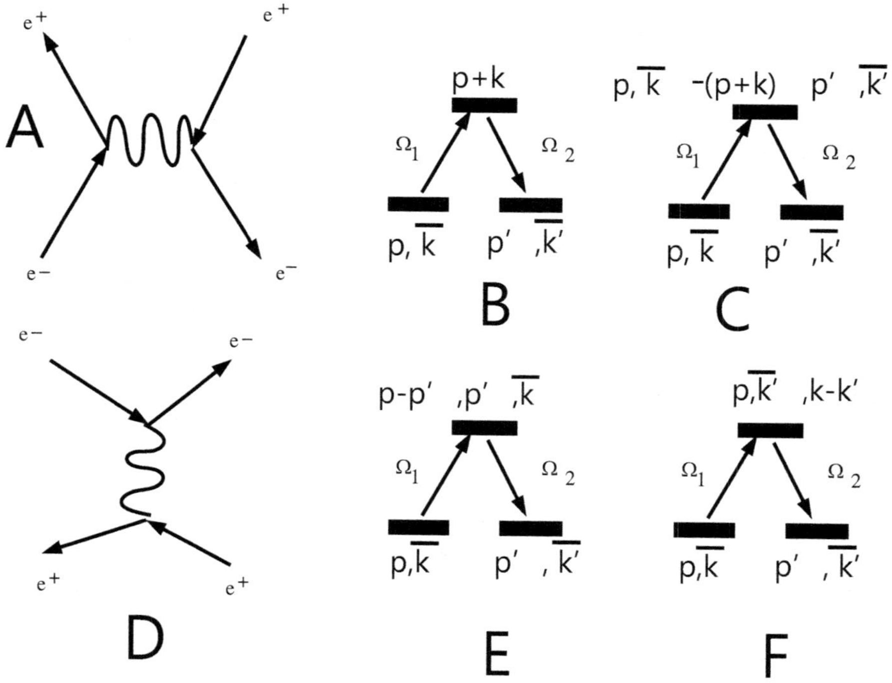

Fig. 3.5 Figure depicts Bhabha scattering. (**a**) is annihilation and (**d**) scattering. (**b**) and (**c**) are three-level processes with (**a**) and (**e**), (**f**) with (**d**)

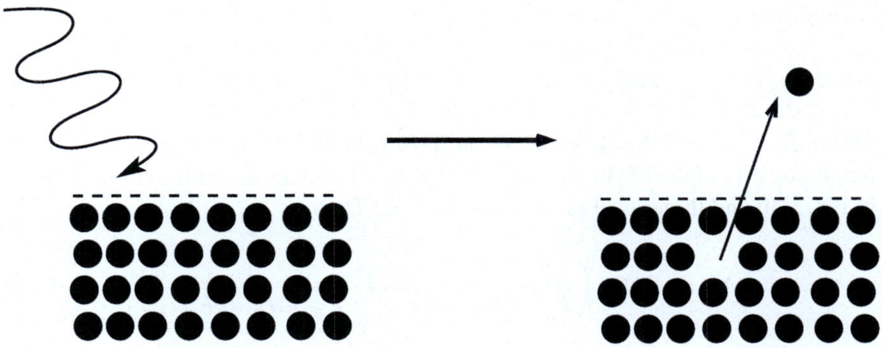

Fig. 3.6 Figure shows how a photon ejects a negative energy electron and creates an electron and positron (hole) pair

$-k$ and missing charge $-e$ or positron (hole) with momentum k and charge e . The process is like photoelectric effect where a valence electron is ejected to a free electron. If we read this process reverse, then we have electron-positron pair of momentum p and k annihilate to form a photon of momentum $p + k$. Denoting electron and positron helicity by s, s' and t, t' etc., the transition amplitude for this process is

$$\Omega_1 = \frac{C}{\sqrt{2E_{p+k}}} \bar{v}_t(k)\epsilon^*_\mu \gamma^\mu u_s(p) \tag{3.29}$$

The photon resulting from annihilation can now do ejection to create electron-positron or electron-hole pair with momentum p' and k', respectively, with transition amplitude:

$$\Omega_2 = \frac{C}{\sqrt{2E_{p+k}}} \bar{u}_{s'}(p')\epsilon_\mu \gamma^\mu v_{t'}(k') \tag{3.30}$$

All this is depicted as a three-level process in Fig. 3.5b. The associated Feynman diagram is in Fig. 3.5a. The energy level difference between the ground and excited states

$$\Delta E_a = E_1 - E_2 = E_p + E_k - E_{p+k}. \tag{3.31}$$

where E_p, E_k, and E_{p+k} are electron, positron, and photon energies.

There is another three-level process in Fig. 3.5c associated with Feynman diagram in Fig. 3.5a. In this process, the negative energy electron just emits a photon with momentum $-(p' + k') = -(p + k)$ and an electron and positron with momentum p' and k'. The amplitude of this process is Ω_2 above. The emitted photon then combines with incoming electron and fills the incoming hole (vacancy) with amplitude same as Ω_1. The energy level difference between the ground and excited states

$$\Delta E'_a = E_1 - E_2 = -(E_p + E_k + E_{p+k}). \tag{3.32}$$

where E_p, E_k, and E_{p+k} are electron, positron, and photon energies. Then, the amplitude $\mathcal{M}_1$ of the Feynman diagram in Fig. 3.5a is the sum of three-level process in Fig. 3.5b and three-level process in Fig. 3.5c. Then

$$\mathcal{M}_a = \Omega_1\Omega_2\left(\frac{1}{E_p + E_k - E_{p+k}} - \frac{1}{E_p + E_k + E_{p+k}}\right) \tag{3.33}$$

$$= \Omega_1\Omega_2\,\frac{2E_{p+q}}{(E_p + E_k)^2 - E^2_{p+k}} \tag{3.34}$$

Now, as in previous section on Moeller scattering, we have

$$\mathcal{M}_F = C^2\,\frac{\bar{v}_t(k)\gamma^\mu u_s(p)\;\;\bar{u}_{s'}(p')\gamma_\mu v_{t'}(k')}{(p+k)^2}. \tag{3.35}$$

3.4.2 Scattering

There is another picture of positron suitable for scattering which is just electron evolving backward, which makes positive energy states as $v(k)$ instead of $u(k)$.

There is one more Feynman diagram in Fig. 3.5d that contributes to this scattering process. There are also two three-level processes Figs. 3.5e and f that contribute to this diagram. In Fig. 3.5e, an electron p scatters to p' giving a photon of momentum $q = p - p'$. This happens with amplitude

$$\Omega_3 = \frac{C}{\sqrt{2E_q}}\;\bar{u}_{s'}(p')\epsilon^*_\nu\gamma^\nu u_s(p) \tag{3.36}$$

and this photon then scatters positron. This happens with amplitude:

$$\Omega_4 = \frac{C}{\sqrt{2E_q}}C\;\bar{v}_{t'}(k')\epsilon_\nu\gamma^\nu v_t(k) \tag{3.37}$$

The difference in energies of ground and excited state is

$$\Delta E_b = E_1 - E_2 = E_p - E_{p'} - E_{p-p'}. \tag{3.38}$$

Positron can emit first and electron can absorb a negative momentum photon. This happens with amplitude:

$$\Omega_3 = \frac{C}{\sqrt{2E_q}}\, \bar{v}_{t'}(k')\epsilon^*_{\nu}\gamma^{\nu} v_t(k) \tag{3.39}$$

This happens with amplitude:

$$\Omega_4 = \frac{C}{\sqrt{2E_q}}\, \bar{u}_{s'}(p')\epsilon_{\nu}\gamma^{\nu} u_s(p) \tag{3.40}$$

$$\Delta E_b = E_1 - E_2 = E_k - E_{k'} - E_{k-k'}. \tag{3.41}$$

Then, the amplitude $\mathcal{M}_2$ of the Feynman diagram in Fig. 3.5d is sum of three-level process in Fig. 3.5e and three-level process in Fig. 3.5f. Then

$$\mathcal{M}_s = \Omega_3\Omega_4 \left(\frac{1}{E_p - E_{p'} - E_{p-p'}} + \frac{1}{E_k - E_{k'} - E_{k-k'}} \right) \tag{3.42}$$

$$= \Omega_3\Omega_4 \left(\frac{1}{E_p - E_{p'} - E_{p-p'}} - \frac{1}{E_p - E_{p'} + E_{p-p'}} \right) \tag{3.43}$$

$$= \Omega_3\Omega_4 \frac{2E_{p-p'}}{(E_p - E_{p'})^2 - E^2_{p-p'}} \tag{3.44}$$

If we denote four momentum, $q = p - p'$, and $q^2 = q_\mu q^\mu$. Now as in previous section on Moeller scattering

$$\mathcal{M}_F = -C^2\, \frac{\bar{u}_{s'}(p')\gamma^{\nu} u_s(p)\ \bar{v}_{t'}(k')\gamma_{\nu} v_t(k)}{(p - p')^2}. \tag{3.45}$$

The total amplitude then is

$$\mathcal{M}_F = C^2 \left(-\frac{\bar{u}_{s'}(p')\gamma^{\nu} u_s(p)\ \bar{v}_{t'}(k')\gamma_{\nu} v_t(k)}{(p-p')^2} + \frac{\bar{v}_t(k)\gamma^{\mu} u_s(p)\ \bar{u}_{s'}(p')\gamma_{\mu} v_{t'}(k')}{(p+k)^2} \right)$$
$$= C^2\mathcal{N}. \tag{3.46}$$

Of course, $\mathcal{M} = \eta\mathcal{M}_F$, the discount factor.

3.4.3 Cross Section

Figure 3.7a shows the schematic of electron-positron each of volume $V = l^3$ colliding head on. We can ask what should be the smallest density or the cross-sectional area $A = l^2$ for the two to scatter at an angle θ with probability 1, when

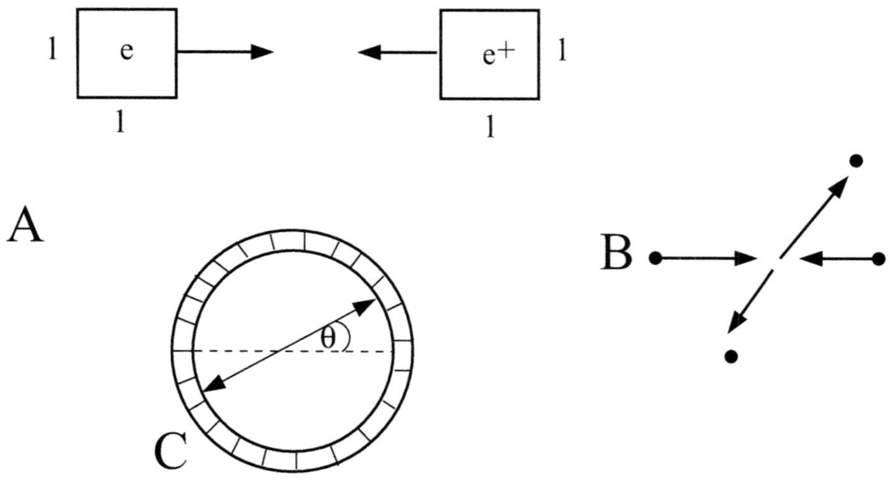

Fig. 3.7 Figure shows the electron-positron colliding and scattering to a different angle

they collide. This is called differential cross section. We have calculated $\mathcal{M}$ the scattering amplitude, in center of mass frame. Let $\mathcal{M}(p_i)$ denote this as function of outgoing momenta p_i. Then, by Fermi golden rule, the probability of scattering P is given by

$$\frac{dP}{dt} = \frac{\sum_i |\mathcal{M}(p_i)|^2}{\hbar \Delta E} \tag{3.47}$$

where ΔE is the energy width of the tessellation of the momentum space volume as shown in Fig. 3.7c. Note $|\mathcal{M}(p_i)|^2$ carries with it a factor C^4 which has in it $\frac{(\hbar c)^4}{V^2}$. Let $\frac{1}{V} = \frac{d^3k}{(2\pi)^3}$ or $\frac{(\hbar c)^3}{V} = \frac{d^3p}{(2\pi)^3}$. With $E = \sqrt{p^2 + m^2}$, we get $\Delta E = \frac{p\Delta p}{E}$. Then, converting sum in 3.47 to integral, we get

$$\frac{dP}{dt} = \frac{e^4 c}{\epsilon_0^2 l^3 (2\pi)^3} \frac{|p|^2 \Delta p \int |\mathcal{N}(p)|^2 d\Omega}{\Delta E} \tag{3.48}$$

Then, it takes $\Delta t = \frac{l}{c}$ for the packets to cross each other and in this time we want

$$\frac{dP}{dt}\frac{l}{c} = 1 \tag{3.49}$$

or we get using ΔE

$$\sigma = l^2 = \frac{e^4}{\epsilon_0^2 (2\pi)^3} E^2 \int |\mathcal{N}(p)|^2 d\Omega \tag{3.50}$$

or

$$\frac{d\sigma}{d\Omega} = \frac{e^4 E^2}{\epsilon_0^2 (2\pi)^3} |\mathcal{N}(p)|^2 \tag{3.51}$$

This is called differential cross section.

Now, we calculate differential cross section by evaluating

$$|\mathcal{N}|^2 = |\mathcal{N}_a + \mathcal{N}_s|^2 \tag{3.52}$$

In fact, we evaluate unpolarized cross section which is to say we average over all possible helicities to get

$$\begin{aligned}\sum_{s,s',t,t'} |\mathcal{N}|^2 &= \sum_{s,s',t,t'} |\mathcal{N}_a + \mathcal{N}_s|^2 \\ &= \sum_{s,s',t,t'} |\mathcal{N}_a|^2 + \sum_{s,s',t,t'} |\mathcal{N}_s|^2 + \sum_{s,s',t,t'} \mathcal{N}_a \mathcal{N}_s^* + \mathcal{N}_a^* \mathcal{N}_s\end{aligned}$$

3.4.4 Relativistic Limit

Let's evaluate the unpolarized cross section:

$$\frac{1}{4} \sum_{s,s',t,t'} |\mathcal{N}_s|^2. \tag{3.53}$$

Recall

$$\mathcal{N}_s = -\frac{\bar{u}_{s'}(p')\gamma^\nu u_s(p)\ \bar{v}_t(k)\gamma_\nu v_{t'}(k')}{(p-p')^2} \tag{3.54}$$

Let electron and positron approach each other along z and $-z$ direction, respectively . Under relativistic limit, helicity 1 electron and positron are

$$u(p) = \begin{bmatrix} 1 \\ 0 \end{bmatrix} \otimes \begin{bmatrix} 1 \\ 0 \end{bmatrix}; \quad v(k) = \begin{bmatrix} 0 \\ 1 \end{bmatrix} \otimes \begin{bmatrix} 0 \\ 1 \end{bmatrix};$$

$$u(p') = \begin{bmatrix} 1 \\ 0 \end{bmatrix} \otimes \begin{bmatrix} \cos\frac{\theta}{2} \\ \sin\frac{\theta}{2} \end{bmatrix}; \quad v(k') = \begin{bmatrix} 0 \\ 1 \end{bmatrix} \otimes \begin{bmatrix} -\sin\frac{\theta}{2} \\ \cos\frac{\theta}{2} \end{bmatrix};$$

Under relativistic limit, helicity -1 electron and positron are

$$u(p) = \begin{bmatrix} 0 \\ 1 \end{bmatrix} \otimes \begin{bmatrix} 0 \\ 1 \end{bmatrix}; \quad v(k) = \begin{bmatrix} 1 \\ 0 \end{bmatrix} \otimes \begin{bmatrix} 1 \\ 0 \end{bmatrix};$$

$$u(p') = \begin{bmatrix} 0 \\ 1 \end{bmatrix} \otimes \begin{bmatrix} -\sin\frac{\theta}{2} \\ \cos\frac{\theta}{2} \end{bmatrix}; \quad v(k') = \begin{bmatrix} 1 \\ 0 \end{bmatrix} \otimes \begin{bmatrix} \cos\frac{\theta}{2} \\ \sin\frac{\theta}{2} \end{bmatrix};$$

Now, observe in 3.54, we get zero if we switch either electron and positron helicity. Furthermore, under relativistic limit, we get

$$(p - p')^2 \sim -2p.p' \sim -2E^2(1 - \cos\theta). \tag{3.55}$$

Then, substituting all helicities in Eq. 3.53, we get

$$\frac{1}{4}\sum_{s,s',t,t'} |\mathcal{N}_s|^2 = \frac{1}{8E^4}.$$

Now, let's evaluate

$$\frac{1}{4}\sum_{s,s',t,t'} |\mathcal{N}_a|^2. \tag{3.56}$$

Recall

$$\mathcal{N}_a = \frac{\bar{v}_t(k)\gamma^\mu u_s(p)\ \bar{u}_{s'}(p')\gamma_\mu v_{t'}(k')}{(p+k)^2} \tag{3.57}$$

Now, observe under relativistic limit, in 3.57, we get zero if incoming or outgoing pair has same helicity. Then, substituting all helicities in Eq. 3.56, we get

$$\frac{1}{4}\sum_{s,s',t,t'} |\mathcal{N}_a|^2 = \frac{(1-\cos\theta)^2}{4E^4}.$$

Finally, evaluating

$$\frac{1}{4}\sum_{s,s',t,t'} \mathcal{N}_a\mathcal{N}_s^* + \mathcal{N}_a^*\mathcal{N}_s = -\frac{(1-\cos\theta)}{4E^4}. \tag{3.58}$$

where we have only two terms, incoming or outgoing pair has same helicity and helicity cannot switch from incoming to outgoing. Adding everything, we get

$$\frac{1}{4}\sum_{s,s',t,t'} |\mathcal{N}|^2 = \frac{(1-\cos\theta)^2 + \cos^2\theta}{8E^4}. \tag{3.59}$$

For $s = E^2$, we get

$$s\frac{d\sigma}{d\Omega} = \frac{e^4}{\epsilon_0^2}\frac{(1-\cos\theta)^2+\cos^2\theta}{32\,\pi^2} \tag{3.60}$$

We can write the cross section as $\frac{d\sigma}{d\Omega} = l^2 f(\theta)$, where $l = \frac{\alpha\hbar c}{E}$. This is a very useful form. The expression

$$l = \frac{\alpha\hbar c}{E}$$

is very aesthetically appealing. At $E = GeV$, we have cross section $l^2 \sim 10^{-8}$ barn, where 1 barn is 10^{-28} m^2.

3.5 Muon Scattering

The electron-positron can annihilate to form muon-anti-muon. We can work out the cross section as $\frac{d\sigma}{d\Omega} = l^2 f(\theta)$, where

$$l = \frac{\alpha\hbar c}{E}\sqrt{1-\frac{(m_\mu c^2)^2}{E^2}}.$$

There is no scattering term in this collision. At $E = GeV$, we have cross section $l^2 \sim 10^{-8}$ barn.

3.6 Compton Scattering

Compton scattering is the inelastic scattering of a photon with an electrically charged particle, first discovered in 1923 by Arthur Compton [6]. This scattering process is of particular historical importance as classical electromagnetism is insufficient to describe the process; a successful description requires us to take into account the particle-like properties of light. Furthermore, the Compton scattering of an electron and a photon is a process that can be described to a high level of precision by QED.

In Compton scattering, an electron and photon with momentum p and k, respectively, scatter into momentum p' and k', respectively. We want to calculate the amplitude for this scattering.

First note with $p \sim 0$, at rest, and k say along x, (see Fig. 3.10), we have the energy of the scattered electron $E = \frac{p'^2}{2m} = \hbar^2\frac{(k-k'\cos\theta)^2+k'^2\sin^2\theta}{2m}$, where θ is angle

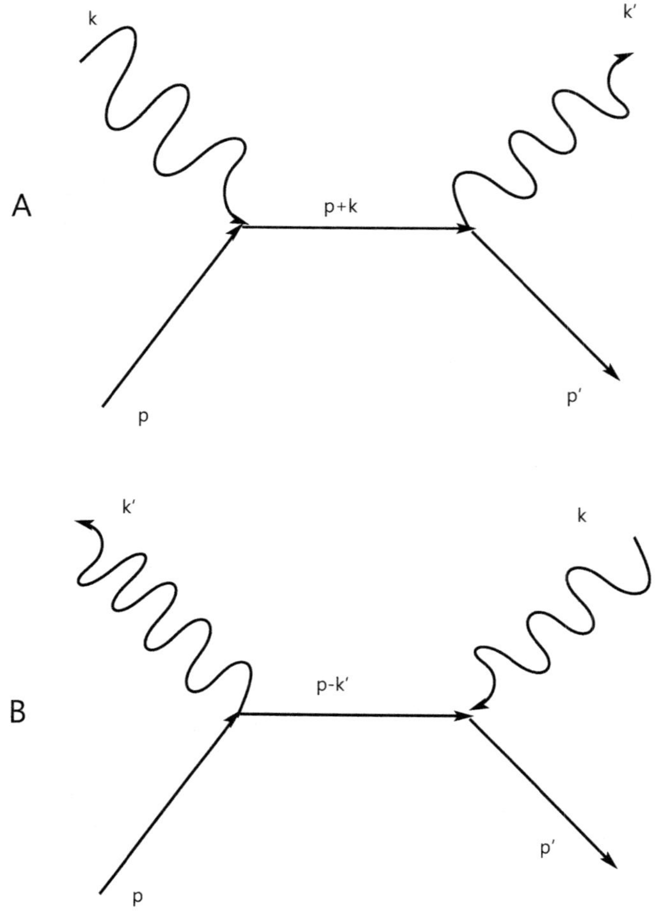

Fig. 3.8 (**a**) and (**b**) show the two mechanisms for Compton scattering

of scattered photon with the x axis. But $E = \hbar c(k - k')$ and we get for $\lambda = \frac{2\pi}{k}$, we have

$$\lambda' - \lambda = \frac{h(1 - \cos\theta)}{mc}$$

There are two Feynman diagrams that show mechanism of Compton scattering. They are shown in Fig. 3.8. We can associate each of these with two three-level diagrams as shown in Fig. 3.9.

Consider Feynman diagram A in Fig. 3.8, where an electron of momentum p and photon of momentum k are annihilated to give an electron of momentum $q = p + k$ which is then annihilated to create electron and photon with momentum p' and k'. This corresponds to three-level system Fig. 3.9a. The scattering amplitude for this

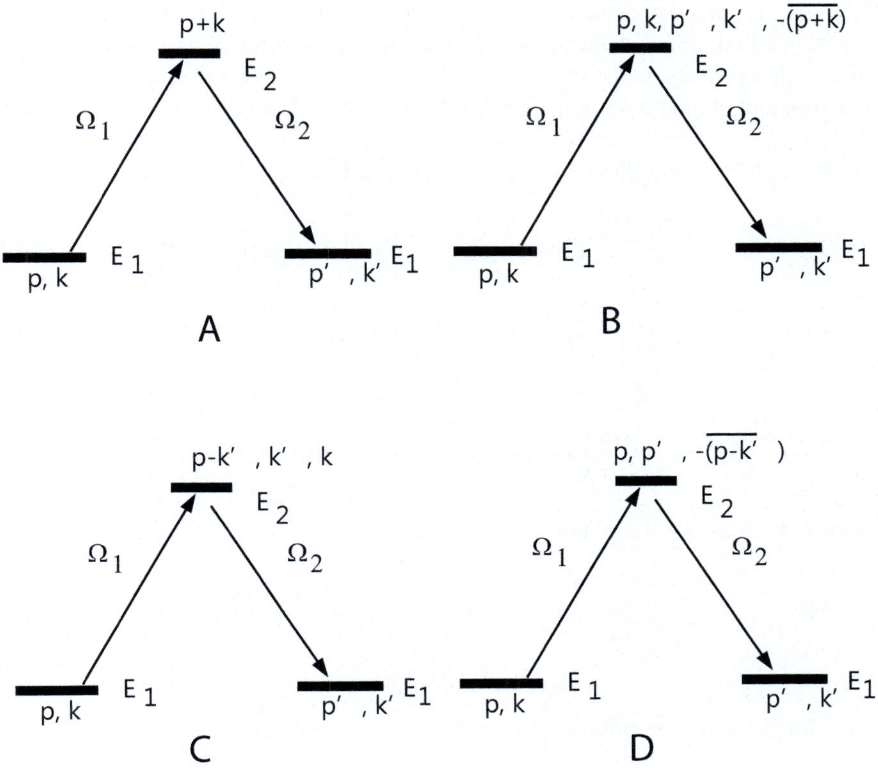

Fig. 3.9 Figure shows three-level systems that go with Feynman diagrams in Fig. 3.8

system is as follows:

$$\Omega_1 = \frac{C}{\sqrt{2E_k}}\,\bar{u}_s(p+k)\gamma^{\nu}\epsilon_{\nu}(k)u(p), \tag{3.61}$$

$$\Omega_2 = \frac{C}{\sqrt{2E_{k'}}}\bar{u}(p')\gamma^{\mu}\epsilon^*_{\mu}(k')u_s(p+k), \tag{3.62}$$

$$E_1 - E_2 = q_0 - E_q = E_p + E_k - E_{p+k}, \tag{3.63}$$

$$\mathcal{M}^s_{1a} = \frac{\Omega_1\Omega_2}{E_1 - E_2}. \tag{3.64}$$

where $E_p = \sqrt{(|p|c)^2 + \mathbf{m}^2}$ and $E_k = |k|c$. Summing over electron polarization, we get

$$\mathcal{M}_{1a} = \frac{C^2}{2\sqrt{E_k E_{k'}}}\,\bar{u}(p')\gamma^{u}\epsilon^*_{\mu}(k')\frac{\sum_s u_s(p+k)\bar{u}_s(p+k)}{E_q(q_0 - E_q)}\gamma^{\nu}\epsilon_{\nu}(k)u(p). \tag{3.65}$$

There is an associated three-level diagram with this as shown in Fig. 3.9b, where we first create electron and photon with momentum p' and k', respectively, alongside a positron with momentum $q = -(p'+k') = -(p+k)$ and then annihilate electron and photon with momentum p and k alongside a positron with momentum $-(p+k)$.

The scattering amplitude for this system is as follows:

$$\Omega_1 = \frac{C}{\sqrt{2E_{k'}}}\, \bar{u}(p')\gamma^{\mu}\epsilon^{*}_{\mu}(k')u_s(p+k), \tag{3.66}$$

$$\Omega_2 = \frac{C}{\sqrt{2E_k}}\bar{u}_s(p+k)\gamma^{\nu}\epsilon_{\nu}(k)u(p), \tag{3.67}$$

$$E_1 - E_2 = -(q_0 + E_q) = -E_{p+k} - (E_p + E_k), \tag{3.68}$$

$$\mathcal{M}^s_{1b} = \frac{\Omega_1\Omega_2}{E_1 - E_2}. \tag{3.69}$$

Summing over electron polarization, we get

$$\mathcal{M}_{1b} = -\frac{C^2}{2\sqrt{E_k E_{k'}}}\, \bar{u}(p')\gamma^{\mu}\epsilon^{*}_{\mu}(k')\frac{\sum_s u_s(p+k)\bar{u}_s(p+k)}{q_0 + E_q}\gamma^{\nu}\epsilon_{\nu}(k)u(p). \tag{3.70}$$

Adding the two amplitudes $\mathcal{M}_1 = \mathcal{M}_{1a} + \mathcal{M}_{1b}$, we get

$$\begin{aligned}\mathcal{M}_1 &= \frac{C^2}{\sqrt{E_k E_{k'}}}\, \frac{\bar{u}(p')\gamma^{\mu}\epsilon^{*}_{\mu}(k')\sum_s u_s(p+k)\bar{u}_s(p+k)\gamma^{\nu}\epsilon_{\nu}(k)u(p)}{q^2 - m_0^2},\\ &= \frac{C^2}{2\sqrt{E_k E_{k'}}}\, \frac{\bar{u}(p')\gamma^{\mu}\epsilon^{*}_{\mu}(k')(\not{q} + m_0)\gamma^{\nu}\epsilon_{\nu}(k)u(p)}{q^2 - m_0^2}.\end{aligned} \tag{3.71}$$

We made use of identity $\sum_s u_s(q)\bar{u}_s(q) = \frac{\not{q}+m_0}{2E_q}$, where $\not{q} = q_j\gamma^j$ (c is implicit). We assume we are in a high energy center of mass frame, which implies $E_p \sim E_k$, and we can write

Now, consider Feynman diagram B in Fig. 3.8, where an electron of momentum p is annihilated and photon of momentum k' is created to give an electron of momentum $q = p - k'$ which is then annihilated along with the photon of momentum k to create electron with momentum p'. This corresponds to three-level system Fig. 3.9C. The scattering amplitude for this system is as follows:

$$\Omega_1 = \frac{C}{\sqrt{2E_{k'}}}\, \bar{u}_s(p-k')\gamma^{\mu}\epsilon^{*}_{\mu}(k')u(p), \tag{3.72}$$

$$\Omega_2 = \frac{C}{\sqrt{2E_k}}\bar{u}(p')\gamma^{\nu}\epsilon_{\nu}(k)u_s(p-k'), \tag{3.73}$$

$$E_1 - E_2 = q_0 - E_q = E_p - E_{p-k'} - E_{k'}, \tag{3.74}$$

$$\mathcal{M}^s_{2a} = \frac{\Omega_1 \Omega_2}{E_1 - E_2}. \tag{3.75}$$

Summing over electron polarization, we get

$$\mathcal{M}_{2a} = \frac{C^2}{2\sqrt{E_k E_{k'}}} \bar{u}(p')\gamma^\nu \epsilon_\nu(k) \frac{\sum_s u_s(p-k')\bar{u}_s(p-k')}{q_0 - E_q} \gamma^\mu \epsilon^*_\mu(k') u(p). \tag{3.76}$$

There is an associated three-level diagram with this as shown in Fig. 3.9d, where we first create electron and annihilate photon with momentum p' and k, respectively, alongside creating a positron with momentum $-(p-k') = -(p'-k)$ and then annihilate electron and create photon with momentum p and k' alongside annihilate positron with momentum $-(p-k')$.

The scattering amplitude for this system is as follows:

$$\Omega_1 = \frac{C}{\sqrt{2E_{k'}}} \bar{u}(p')\gamma^\nu \epsilon_\nu(k) u_s(p'-k), \tag{3.77}$$

$$\Omega_2 = \frac{C}{\sqrt{2E_k}} \bar{u}_s(p-k')\gamma^\mu \epsilon^*_\mu(k') u(p), \tag{3.78}$$

$$E_1 - E_2 = -(q_0 + E_q) = -E_{p-k'} - E_p + E_{k'}, \tag{3.79}$$

$$\mathcal{M}^s_{2b} = \frac{\Omega_1 \Omega_2}{E_1 - E_2}. \tag{3.80}$$

Summing over electron polarization, we get

$$\mathcal{M}_{2b} = -\frac{C^2}{2\sqrt{E_k E_{k'}}} \bar{u}(p')\gamma^\nu \epsilon_\nu(k) \frac{\sum_s u_s(p-k')\bar{u}_s(p-k')}{q_0 + E_q} \gamma^\mu \epsilon^*_\mu(k') u(p). \tag{3.81}$$

Adding the two amplitudes $\mathcal{M}_2 = \mathcal{M}_{2a} + \mathcal{M}_{2b}$, we get

$$\begin{aligned}\mathcal{M}_2 &= \frac{C^2}{\sqrt{E_k E_{k'}}} \bar{u}(p')\gamma^\nu \epsilon_\nu(k) \frac{\sum_s u_s(p-k')\bar{u}_s(p-k')}{q^2 - m_0^2} \gamma^\mu \epsilon^*_\mu(k') u(p), \\ &= \frac{C^2}{2\sqrt{E_k E_{k'}}} \frac{\bar{u}(p')\gamma^\nu \epsilon_\nu(k)(\not{q} + m_0)\gamma^\mu \epsilon^*_\mu(k') u(p)}{q^2 - m_0^2}.\end{aligned} \tag{3.82}$$

$$\mathcal{M}_F = \mathcal{M}_1 + \mathcal{M}_2 \tag{3.83}$$

Of course, $\mathcal{M} = \eta \mathcal{M}_F$, the discount factor.

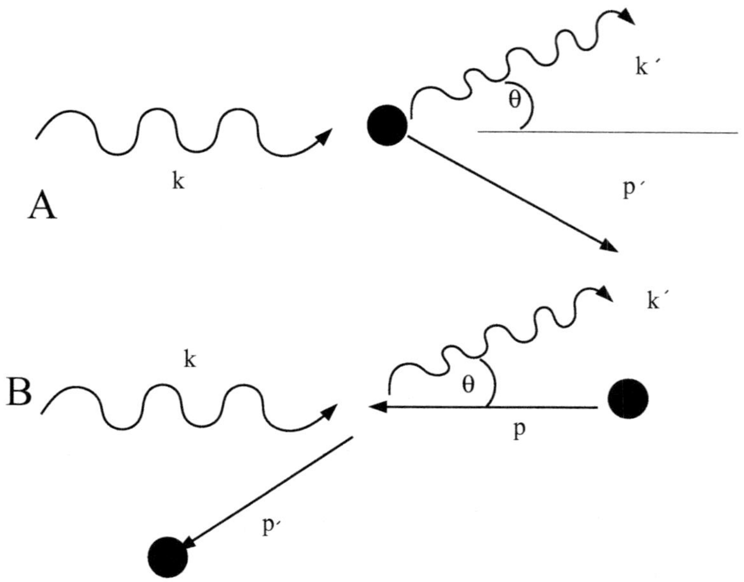

Fig. 3.10 (**a**) shows Compton scattering in lab frame. (**b**) shows Compton scattering in center of mass frame

3.6.1 *Cross Section*

We have to calculate $|\mathcal{M}|^2$ to find cross section. Before we do this, we can just say that we work in regime (as in original Compton's experiment) where photon energy (10s KeV, wavelength A°) is smaller than rest energy of the electron (Mev, Compton wavelength, .01 A°); we have the cross section of the form $\frac{d\sigma}{d\Omega} = l^2 f(\theta)$, where

$$l = \frac{\alpha \hbar c}{E}.$$

Calculating $f(\theta)$ is just an exercise where you just have to roll your sleeves. At $E = 10$ keV, we have cross section ~ 100 barn (Fig. 3.10).

3.7 Vacuum Polarization

Quantum electrodynamics (QED) is one of the most successful theories of modern physics era [2–5]. In QED, electrons interact by electromagnetic coupling to vacuum. Electron emits photon which is absorbed by the second electron leading to momentum exchange between electrons which we call electric force. The emission and absorption changes the energy of the two electrons by what we call the electric

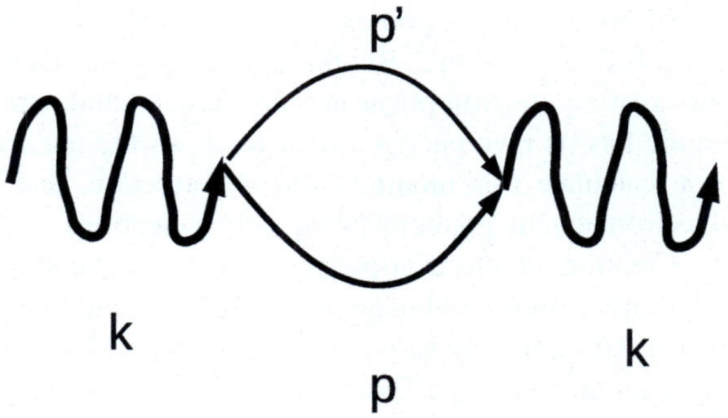

Fig. 3.11 Figure depicts vacuum polarization. Emitted photon generates electron-positron pair which recombine to give the photon back

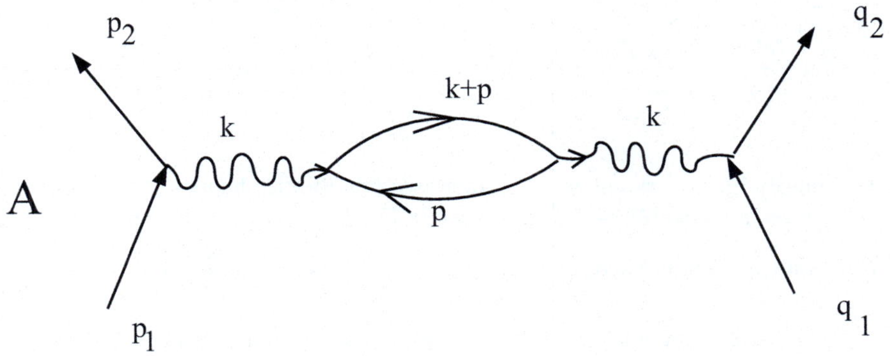

Fig. 3.12 Figure A shows Feynman diagram for vacuum polarization

potential energy. In calculating this energy, which is a second-order calculation, we make use of the energy of photon $E_k = \hbar ck$ where k is its momentum. But this emitted photon can further interact with the vacuum by creating electron-positron pairs, which annihilate to give the photon back. This again has its own energy which modifies the energy of the photon E_k to E'_k. We can calculate this modification or correction, and we find this will change the electromagnetic potential between two electrons. We may think of this as simply changing ϵ_0 the vacuum permittivity, and this is called vacuum polarization, very much like light propagating in a medium polarizes it and changes ϵ_0 and slows down. On another note, an electron can emit and absorb a photon and the process modifies the rest energy of the electron mc^2 to $m'c^2$ a process we call mass correction (Fig. 3.11).

But there is a problem in QED. When we calculate these corrections, we find them divergent. There is a huge body of work in field of QED, that tries to tame these infinities, a process we call renormalization [4]. But, where is the problem? The problem is when we calculate the modification of photon energy we collide it with a sea electron, but the collision is not in center of mass frame; to get the right amplitude, we have to use the correct discount factor η, and then, we find our answers are finite (Fig. 3.12).

Electron 1 emits photon with momentum k and energy $E_k = \hbar ck$, which is absorbed by electron 2. But this emitted photon can further interact with the vacuum by creating electron-positron pairs, which annihilate to give the photon back. This again has its own energy which modifies the energy of the photon E_k to E'_k. We can calculate this modification or correction, and we find this will change the electromagnetic potential between two electrons.

Creation of electron-positron pair is tantamount to Compton scattering; the photon k collides with negative energy, sea, electron with momentum p and energy $-E_p$ and creates a positive energy electron with momentum $p' = p+k$ and positron with momentum $-p$. The name of the game is to sum the amplitude of the process for large values of p.

$$\mathcal{M} = \frac{(e\hbar c)^2}{\epsilon_0 V} \frac{E_{p'} + E_p}{E_k} \left(\bar{v}(-p)\, \gamma^{\mu} u(p')\bar{u}(p')\gamma_{\mu}\, v(-p)\right) \frac{1}{(E_{p'} + E_p)^2 - E_k^2} \tag{3.84}$$

Simplify by $k \sim 0$ and $p \sim p'$, then two possibilities happen, one when spin of electron and positron are aligned, then in Eq. 3.84, $\mathcal{M} \sim 0$, second possibility when they are antialligned, then for large p, we have one as $\begin{pmatrix} 1 \\ 0 \end{pmatrix} \otimes \uparrow$ and other $\begin{pmatrix} 0 \\ 1 \end{pmatrix} \otimes \downarrow$ and again in Eq. 3.84, $\mathcal{M} \sim 0$. We can show spinor part of Eq. 3.84 goes as $\sim \frac{1}{|p|^2}$. Now, taking other factors into account and the discount factor, $\eta \sim \frac{1}{|p|}$, we have

$$\mathcal{M} \sim \frac{1}{E_p^4} = \frac{1}{|p|^4},$$

Summing over $|p|$, we get the harmonic sum $\int \frac{1}{|p|^2} dp$, which nicely converges.

3.8 Electron Self-Energy

In the last section, we talked about vacuum polarization, where a photon splits into an electron-positron pair and recombines. In this section, we discuss another QED process, the electron self-energy. Hereby, an electron of momentum p emits a photon and then reabsorbs it. This is shown in Fig. 3.13a. This process can be represented by two-level diagrams as in Fig. 3.13b. In the first one, we have an electron with momentum p emit a photon with momentum k and subsequently reabsorb it. In the second one, we have creation of a positron, electron, and photon with momentum $-p$, $p - k$, and k, respectively, and their subsequent annihilation.

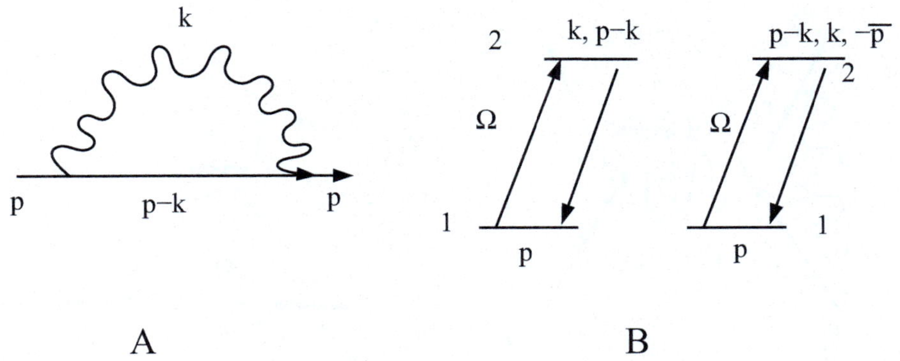

Fig. 3.13 (**a**) shows corrections to electron energy, where an electron emits and absorbs a photon. (**b**) shows two-level diagrams for this process

Electron emits photon, changes momentum $p' = p - k$, and reabsorbs photon to get to its initial state. The name of the game is to sum the amplitude of the process for large values of k.

$$\mathcal{M} = \frac{(\hbar c)^2}{\epsilon_0 V} \frac{E_{p'} + E_k}{E_k} \frac{\bar{u}(p)\gamma^{\mu} u(p')\, u(p')\gamma_{\mu}\, u(p)}{(E_k + E'_p)^2 - E_p^2} \tag{3.85}$$

Simplify, choosing $p \sim 0$ and $p' \sim -k$, since p nonrelativistic its spin can be chosen aligned with p' and then for large k the spinor part of 3.85 $\sim \frac{1}{|k|}$ and with discount $\eta \sim \frac{1}{|k|}$, and the whole $\mathcal{M} \sim \frac{1}{|k|^4}$. Summing over $|k|$, we get the harmonic sum $\int \frac{1}{|k|^2} dk$, which nicely converges.

The self-energy process may be thought of as a collision between electron and negative energy photon in opposite direction, justifying use of η.

3.9 Vertex Corrections

Consider the Feynman diagram in Fig. 3.14a. It shows Moller scattering of incoming electron with momentum p_1 and a heavy particle with momentum r_1. Incoming electron emits a photon with momentum k that recombines with outgoing electron with momentum p_2. Figure B shows an equivalent five-level system. The incoming particles with momentum p_1, r_1 are at level 1. Emission of a photon with momentum k transits to level 2. Level 2, 3, 4 represent the Moller scattering of electron and particle with momentum $p_1 - k$ and r_1 to momentum $p_2 - k$ and r_2, and finally, the emitted photon k is reabsorbed, and we get to level 5 with outgoing particles with momentum p_2, r_2.

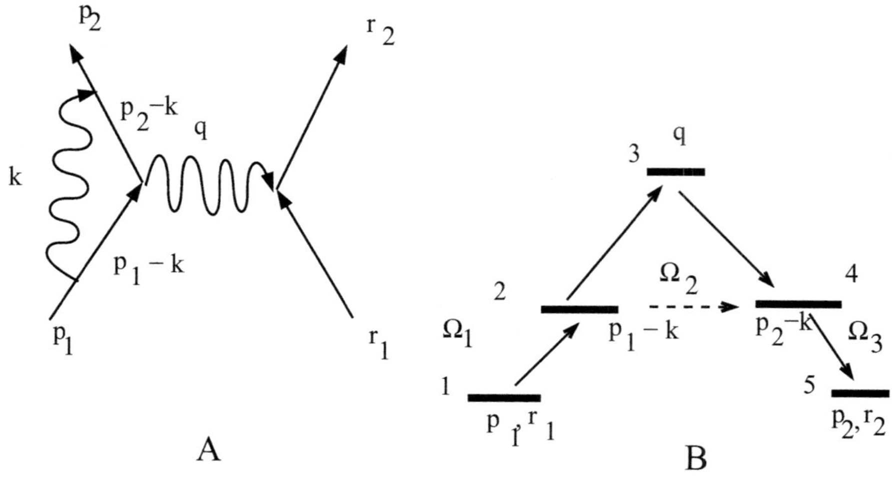

Fig. 3.14 (**a**) shows Moller scattering of incoming electron with momentum p_1 and a heavy particle with momentum p_2. Incoming electron emits a photon with momentum k that recombines with outgoing electron with momentum $p_2 - k$. (**b**) shows an equivalent five-level system. The incoming particles with momentum p_1, r_1 are at level 1. Emission of a photon with momentum k transits to level 2. Level 2, 3, 4 represent the Moller scattering and finally the emitted photon k is reabsorbed and we get to level 5

Let's calculate the scattering amplitude of p_1, r_1 to p_2, r_2 and in the process calculate the new transition amplitude of scattering from p_1 to p_2. This modification of amplitude of scattering from p_1 to p_2 as compared to one studied in Sect. 3.2 is called the *vertex correction*.

Observe under nonrelativistic limit

$$E_{12} = E_1 - E_2 = E_{p_1} - (E_{p_1-k} + E_k) \sim E_{p_2} - (E_{p_2-k} + E_k) = E_{45} \quad (3.86)$$

Then, the transition amplitude from level 1 to level 5 is a third-order term and simply (see the end of the section)

$$\mathcal{M}_F = \frac{\Omega_1\Omega_2\Omega_3}{E_{12}^2}, \quad (3.87)$$

where Ω_i are as in Fig. 3.14b.

$$\mathcal{M} = \frac{\Omega_1\Omega_2\Omega_3}{(E_{p_1-k} + E_k)^2 - E_{p_1}^2} \frac{E_{p_1-k} + E_k + E_{p_1}}{E_{p_1-k} + E_k - E_{p_1}}, \quad (3.88)$$

where

$$\Omega_1 = \frac{C}{\sqrt{2E_k}} \bar{u}(p_1 - k)\gamma^\nu \epsilon_\nu^*(k) u(p_1), \quad (3.89)$$

$$\Omega_2 \propto C^2 \frac{\bar{u}(r_2)\gamma^\mu \epsilon_\mu(q) u(r_1)\, \bar{u}(p_2 - k)\gamma^\mu \epsilon^*_\mu(q) u(p_1 - k)}{q^2}, \tag{3.90}$$

$$\Omega_3 = \frac{C}{\sqrt{2E_k}}\, \bar{u}(p_2)\gamma^\nu \epsilon(k) u(p_2 - k). \tag{3.91}$$

For large k, using discount $\eta \sim \frac{1}{|k|}$, we get $\mathcal{M} = \mathcal{M}_F \eta \sim \frac{1}{|k|^4}$, and hence summing over $|k|$, we get the harmonic sum $\int \frac{1}{|k|^2} dk$, which nicely converges.

We end the section by sketching the proof for Eq. 3.87. The state of the four-level system (level 1, 2, 4, 5 in Fig. 3.14b) evolves according to the Schrödinger equation:

$$\dot{\psi} = \frac{-i}{\hbar} \begin{bmatrix} E_1 & \Omega_1^* & 0 & 0 \\ \Omega_1 & E_2 & \Omega_2^* & 0 \\ 0 & \Omega_2 & E_2 & \Omega_3^* \\ 0 & 0 & \Omega_3 & E_1 \end{bmatrix} \psi. \tag{3.92}$$

We proceed into the interaction frame of the natural Hamiltonian (system energies) by transformation:

$$\phi = \exp(\frac{i\,t}{\hbar} \begin{bmatrix} E_1 & 0 & 0 & 0 \\ 0 & E_2 & 0 & 0 \\ 0 & 0 & E_2 & 0 \\ 0 & 0 & 0 & E_1 \end{bmatrix})\psi. \tag{3.93}$$

This gives for $E_{12} = E_2 - E_1$

$$\dot{\phi} = \underbrace{\frac{-i}{\hbar} \begin{bmatrix} 0 & \exp(-\frac{i}{\hbar} E_{12}\, t)\Omega_1^* & 0 & 0 \\ \exp(\frac{i}{\hbar} E_{12}\, t)\Omega_1 & 0 & \Omega_2^* & 0 \\ 0 & \Omega_2 & 0 & \exp(\frac{i}{\hbar} E_{12}\, t)\Omega_3^* \\ 0 & 0 & \exp(\frac{-i}{\hbar} E_{12}\, t)\Omega_3 & 0 \end{bmatrix}}_{H(t)} \phi. \tag{3.94}$$

$H(t)$ is periodic with period $\Delta t = \frac{2\pi}{E_{12}}$. After Δt, the system evolution is $\phi(\Delta t) =$

$$(I + \int_0^{\Delta t} H(\sigma) d\sigma + \int_0^{\Delta t} \int_0^{\sigma_1} H(\sigma_1) H(\sigma_2) d\sigma_2 d\sigma_1 + \int_0^{\Delta t} \int_0^{\sigma_1} \int_0^{\sigma_2} H(\sigma_1) H(\sigma_2) H(\sigma_3) d\sigma_3 d\sigma_2 d\sigma_1 \ldots)\phi(0). \tag{3.95}$$

The first integral averages to zero, while the second integral doesn't give transition between 1 and 4. The third order does with a contribution:

$$\int_0^{\Delta t} \Omega_1 \exp(\frac{i}{\hbar} E_{12}\sigma_1) \int_0^{\sigma_1} \Omega_2 \int_0^{\sigma_2} \Omega_3 \exp(\frac{-i}{\hbar} E_{12}\,\sigma_3) d\sigma_3 d\sigma_2 d\sigma_1 = 2\Delta t \frac{\Omega_1\Omega_2\Omega_3}{E_{12}^2}.$$

3.10 Lamb Shift

This vertex correction accounts for slight energy shift ~ 1 GHz between $2S_{\frac{1}{2}}$ and $2P_{\frac{1}{2}}$ energy levels of hydrogen atom. It was discovered by Willis Lamb in 1951.

3.11 Anomalous Magnetic Moment of Electron

Figure 3.15 shows the origin to anomalous magnetic moment. Electron with momentum p makes a emission of photon with momentum k and transits to $p - k$ where it spin flips due to Rabi photon and becomes $q - k$, and then, it reabsorbs the emitted photon to return to state q with flipped spin. The spin flip frequency is the Rabi frequency Ω_1 which is modified to photon emission process, and we can say the gyromagnetic ration $\gamma \rightarrow \gamma'$, where $\gamma = g\frac{e}{2m}$, or we can say g changes. To find the change in Ω_1, we have to write the effective transition frequency which is

$$\Omega_{eff} = \frac{\Omega\Omega_1\Omega}{E_k^2},$$

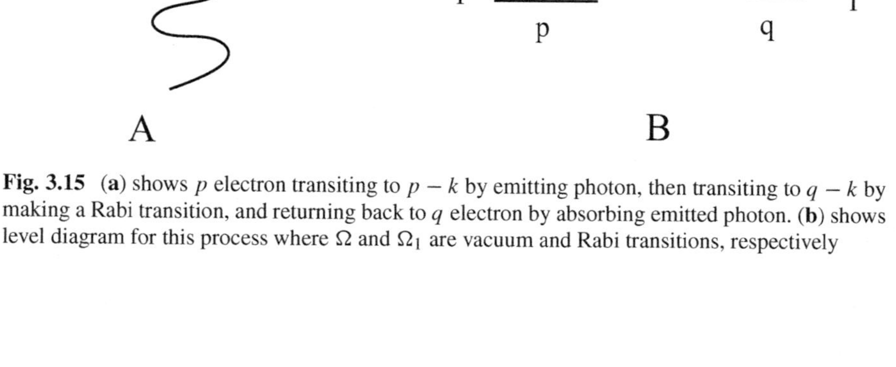

Fig. 3.15 (**a**) shows p electron transiting to $p - k$ by emitting photon, then transiting to $q - k$ by making a Rabi transition, and returning back to q electron by absorbing emitted photon. (**b**) shows level diagram for this process where Ω and Ω_1 are vacuum and Rabi transitions, respectively

but $\Omega \sim \frac{1}{\sqrt{E_k}}$ and we have the discount $\eta \sim \frac{1}{E_k}$ which gives $\Omega_{eff} = \frac{1}{E_k^4} \propto \frac{1}{|k|^4}$. Summing over $|k|$, we get the harmonic sum $\int \frac{1}{|k|^2} dk$, which nicely converges.

3.12 Problems

1. Matrices A, B anticommute, if $AB + BA = 0$. Show that Dirac matrices $\alpha_1, \alpha_2, \alpha_3, \beta$ all anticommute.
2. An electron velocity is $(v_x, v_y, v_z) = v(\sin\theta, 0, \cos\theta)$. If $v = .9c$, find the two electron spinors with positive energy.
3. In the above problem, find the two electron spinors with negative energy.
4. An electron with energy E travelling along z direction collides with an positron with energy E travelling along $-z$ direction. If $E \gg m_e c^2$, (m_e is rest mass of electron), find the differential cross section $\frac{d\sigma}{d\Omega}$.
5. A photon with wavelength λ moving along z direction collides with an electron at rest and starts traveling with velocity $c(\sin\theta, 0, \cos\theta)$; find the new wavelength λ' of the photon.

References

1. W.A. Cottingham, D.A. Greenwood, *An Introduction to Standard Model of Particle Physics* (Cambridge Univ. Press, 2007)
2. M.E. Peskin, D.V. Schroeder, *An Introduction to Quantum Field Theory* (Westview Press, 1995)
3. D. Griffiths, *Introduction to Elementary Particles* (Wiley-VCH, 2004)
4. M. Schwartz, *Quantum Field Theory and the Standard Model* (Cambridge University Press, 2014)
5. M. Thomson, *Modern Particle Physics* (Cambridge University Press, 2013)
6. A.H. Compton, A quantum theory of the scattering of x-rays by light elements. Phys. Rev. **21**(5), 483–502 (1923)
7. D.J. Griffiths, *Introduction to Quantum Mechanics* (Pearson Prentice Hall, 2004)

Chapter 4
Weak Interactions

4.1 Massive Fields

EM photon is not the only photon. EM vacuum is not the only vacuum. We have other photons that mediate the so-called weak interactions as these photons are heavy and we call them W-Z bosons. In this chapter, we develop the theory of W-Z bosons and interactions they mediate.

We equip massive A with a dynamics by defining Lagrangian as density:

$$L = \epsilon_0 \left(-\frac{1}{4} F_{\mu\nu} F^{\mu\nu} + \frac{1}{2} (\frac{\mathbf{m}c}{\hbar})^2 A_\mu A^\mu \right). \tag{4.1}$$

We just write

$$L = -\frac{1}{4} F_{\mu\nu} F^{\mu\nu} + \frac{1}{2} m^2 A_\mu A^\mu. \tag{4.2}$$

where recall

$$F_{\mu\nu} = \partial_\mu A_\nu - \partial_\nu A_\mu \tag{4.3}$$

The energy density of this field is

$$H = -F_{0\mu} F^{0\mu} + \frac{1}{4} F_{\mu\nu} F^{\mu\nu} - \frac{1}{2} m^2 A_\mu A^\mu. \tag{4.4}$$

Variation of L gives

$$\partial_\mu F^{\mu\nu} + m^2 A^\nu = 0 \tag{4.5}$$

$$\partial_\mu \partial^\mu A^\nu - \partial^\nu (\partial_\mu A^\mu) + m^2 A^\nu = 0 \tag{4.6}$$

N. Khaneja, *High Energy Physics*, https://doi.org/10.1007/978-3-031-75645-0_4

Observe

$$\partial_{\mu\nu} F^{\mu\nu} = 0 \tag{4.7}$$

which gives

$$\partial_\mu A^\mu = 0 \tag{4.8}$$

$$\partial_\mu \partial^\mu A^\nu + m^2 A^\nu = 0 \tag{4.9}$$

or

$$\left(\frac{\partial^2}{c^2 \partial t^2} - \nabla^2 + m^2 \right) A^\nu = 0 \tag{4.10}$$

Solution is $\epsilon \exp(j(kx - \omega t))$, where

$$k_0 = \frac{\omega}{c} = \sqrt{k^2 + m^2} \tag{4.11}$$

$$\partial_\mu A^\mu = 0 \rightarrow k_\mu \epsilon^\mu = 0 \tag{4.12}$$

Consider field in z direction. There are three independent polarization directions:

$$\varepsilon_1 = (0, 1, 0, 0) \tag{4.13}$$

$$\varepsilon_2 = (0, 0, 1, 0) \tag{4.14}$$

$$\varepsilon_3 = \frac{1}{m}(k, 0, 0, k_0) \tag{4.15}$$

For example, consider a massive photon:

$$A\varepsilon_{1,2} \cos(k \cdot z - \omega t), \tag{4.16}$$

propagating in z direction with $\frac{\omega}{c} = \sqrt{k^2 + m^2}$. From 4.4, the energy of this photon is $\frac{\epsilon_0 A^2 \omega^2}{2c^2} V$. Therefore, for $\frac{\epsilon_0 A^2 \omega^2}{2c^2} V = \hbar\omega$, we have the photon:

$$\begin{aligned} A &= c\sqrt{\frac{2\hbar}{V\epsilon_0\omega}} \varepsilon_{1,2} \cos(k \cdot z - \omega t) \\ &= c\sqrt{\frac{\hbar}{2\epsilon_0 \omega V}} \varepsilon_{1,2} \left(\exp\ i(k \cdot z - \omega t) + \exp\ -i(k \cdot z - \omega t) \right). \end{aligned} \tag{4.17}$$

Consider the massive photon:

$$A\varepsilon_3 \cos(k \cdot z - \omega t), \tag{4.18}$$

propagating in z direction. The energy of this photon is $\frac{\epsilon_0 A^2 m^2}{2} V$. Therefore, for $\frac{\epsilon_0 A^2 m^2}{2} V = \hbar\omega$, we have the photon:

$$A = \sqrt{\frac{2\hbar\omega}{V\epsilon_0 m^2}}\varepsilon_3 \cos(k \cdot z - \omega t) \sim c\sqrt{\frac{\hbar}{2V\epsilon_0\omega}}\varepsilon_3 \; \cos(k \cdot z - \omega t). \tag{4.19}$$

where the last approximation is true when $k << m$.

4.2 Charged Weak Interaction

There are two charged massive bosons that mediate charged weak interaction. The boson W^+ with momentum k takes in a electron of momentum p and emits a neutrino of momentum $p + k$ as shown in Fig. 4.1a. The amplitude for the transition is

$$\Omega = \frac{C}{\sqrt{2m}}\bar{u_2}(p + k)\bar{\gamma}^{\nu}\epsilon_{\nu}(k)u_1(p), \tag{4.20}$$

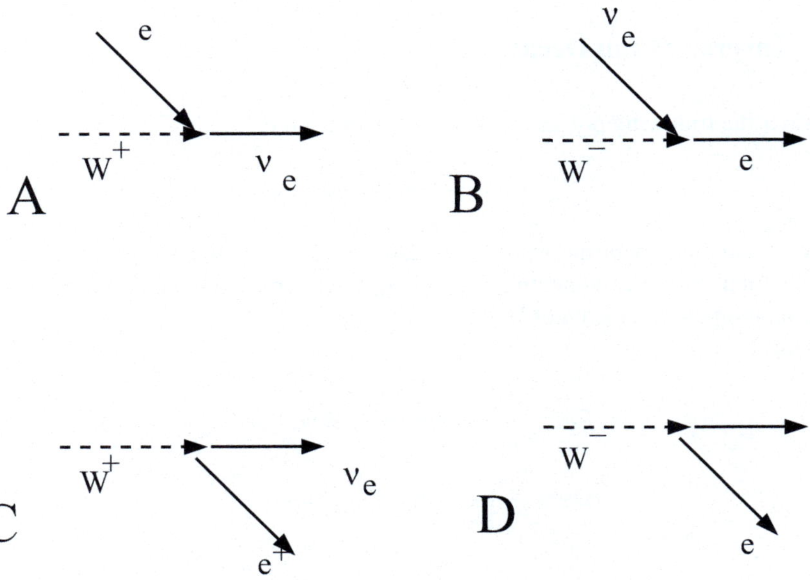

Fig. 4.1 Figure shows vertices for charged weak interactions

where $C = \frac{\hbar c g_w}{\sqrt{V}}$. Here, g_w is weak coupling constant and analogous to $\frac{e}{\sqrt{\epsilon_0}}$ in QED, and

$$\bar{\gamma}^\nu = \gamma^\nu \begin{pmatrix} \mathbf{1} & 0 \\ 0 & 0 \end{pmatrix} = \gamma^\nu \frac{1-\gamma^5}{2} \tag{4.21}$$

where $\gamma^5 = i\gamma^0\gamma^1\gamma^2\gamma^3$. This ensures only left ψ_L of the spinor $\psi = \begin{bmatrix} \psi_L \\ \psi_R \end{bmatrix}$ takes part in weak interaction. This is called a V-A vertex of weak interaction and arises from parity violation in weak interaction as explained subsequently.

In Fig. 4.1c, the boson W^+ with momentum k emits a positron of momentum $-p$ and emits a neutrino of momentum $p + k$ as shown in Fig. 4.1a. The amplitude for the transition is

$$\Omega = \frac{C}{\sqrt{2m}} \bar{u_2}(p+k)\bar{\gamma}^\nu \epsilon_\nu(k) v_1(p). \tag{4.22}$$

In Fig. 4.1b, we consider boson W^- instead of W^+. The boson W^- with momentum k takes in a neutrino of momentum p and emits a neutrino of momentum $p + k$ as shown in Fig. 4.1a. The amplitude for the transition is

$$\Omega = \frac{C}{\sqrt{2m}} \bar{u_2}(p+k)\bar{\gamma}^\nu \epsilon_\nu(k) u_1(p). \tag{4.23}$$

4.3 Inverse Muon Decay

Consider the following process mediated by weak force:

$$e + \nu_\mu \rightarrow \nu_e + \mu \tag{4.24}$$

Electron and muon neutrino with momentum p_1 and p_2 collide to produce electron neutrino and muon at momentum p_3 and p_4. Let k and q denote the on-shell and off-shell momenta of mediator W boson (Fig. 4.2).

With

$$\Omega_1 = \frac{C}{\sqrt{2m}} \bar{u}(p_3)\bar{\gamma}^\nu \epsilon_\nu^*(k) u(p_1) \tag{4.25}$$

$$\Omega_2 = \frac{C}{\sqrt{2m}} \bar{u}(p_4)\bar{\gamma}^\mu \epsilon_\mu(k) u(p_2). \tag{4.26}$$

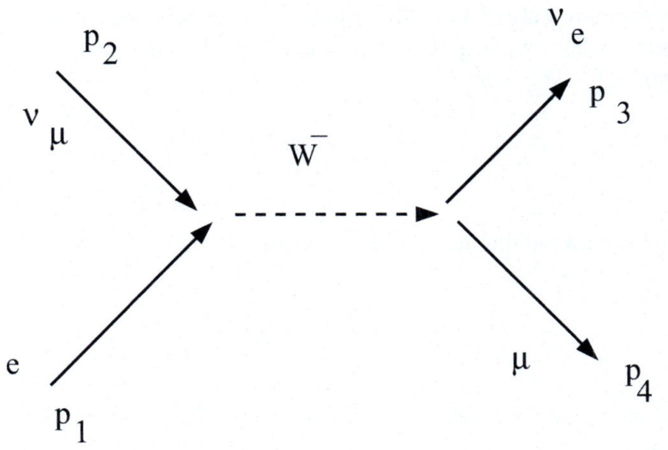

Fig. 4.2 Figure shows inverse muon decay $e + \nu_\mu \rightarrow \mu + \nu_e$

The amplitude for the process

$$\begin{aligned}
\mathcal{M} &= \Omega_1\Omega_2 \left(\frac{1}{E_e(p_1) + E_{\nu_e}(p_3) - E_{W^-}(q)} - \frac{1}{E_{W^+}(-q) + E_\mu(p_4) - E_{\mu_\nu}(p_2)} \right) \\
&= \Omega_1\Omega_2 \left(\frac{1}{E_e(p_1) + E_{\nu_e}(p_3) - E_{W^-}(q)} - \frac{1}{E_{W^+}(-q) + E_e(p_1) - E_{\nu_e}(p_3)} \right) \\
&= \Omega_1\Omega_2 \left(\frac{2E_{W^-}(q)}{q^2 - m_W^2} \right) \\
&\sim \frac{C^2}{q^2 - m_W^2}\, \bar{u}(p_4)\bar{\gamma}^\mu \epsilon_\mu(k) u(p_2)\, \bar{u}(p_3)\bar{\gamma}^\nu \epsilon_\nu^*(k) u(p_1)
\end{aligned}$$

Now, we have to sum over the polarization ϵ. Lorentz invariance arguments given in QED dictate that amplitude $\mathcal{M}$ be

$$\mathcal{M} \sim \frac{C^2}{m_W^2} \bar{u}(p_4)\bar{\gamma}^\mu u(p_2)\, \bar{u}(p_3)\bar{\gamma}_\mu u(p_1) \tag{4.27}$$

$$= \frac{C^2}{4m_W^2} \bar{u}(p_4)\gamma^\mu(1-\gamma^5)u(p_2)\, \bar{u}(p_3)\gamma_\mu(1-\gamma^5)u(p_1). \tag{4.28}$$

where we use approximation $q^2 \ll m_W^2$.

Not worrying too much about the spinor part, we can just write the cross section in CM frame, where we neglect electron mass; we find $E_1 = E_2 = E$ and $E_3 \sim E_4 = E$ and $E_4^2 - E_3^2 = m_\mu^2$:

$$l = \frac{\alpha_w \hbar c E}{m_W^2}$$

or more precise taking muon mass into account

$$l = \frac{\hbar c E}{m_W^2}(1 - \frac{m_\mu^2}{E^2}).$$

For collisions at Gev, we get cross section $\sim 10^{-14}$ barn. For those interested in spinor part

$$\begin{aligned}\sum_s |\mathcal{M}|^2 = (\tfrac{C^2}{16 m_W^2})^2 \tfrac{1}{E_1 E_2 E_3 E_4} Tr(\not{p}_3 \gamma^\mu (1-\gamma^5)(\not{p}_1 + m_e)\gamma^\nu (1-\gamma^5)) \\ \times Tr((\not{p}_4 + m_\mu)\gamma_\mu (1-\gamma^5)\not{p}_2 \gamma_\nu (1-\gamma^5)).\end{aligned} \tag{4.29}$$

With lot of algebra

$$\sum_s |\mathcal{M}|^2 = (\frac{C^2}{m_W^2})^2 \frac{(p_1 \cdot p_2)\,(p_3 \cdot p_4)}{E_1 E_2 E_3 E_4} \tag{4.30}$$

In the CM frame where we neglect electron mass, we find $E_1 = E_2 = E$ and $E_3 \sim E_4 = E$ and $E_4^2 - E_3^2 = m_\mu^2$. $E \sim \sqrt{|\mathbf{p}|^2 + m_\mu^2/2}$, where p is the muon momentum:

$$\sum_s |\mathcal{M}|^2 = (\frac{2C^2}{m_W^2})^2 (1 - \frac{m_\mu^2}{2E^2}) \tag{4.31}$$

$$\frac{d\sigma}{d\Omega} = \left(\frac{2\alpha_w}{m_W^2}\,\hbar c\; E\;(1 - \frac{m_\mu^2}{2E^2})\right)^2 \tag{4.32}$$

4.4 Muon Decay

Figure 4.4a shows the decay of a muon where the amplitude of the Feynman diagram is in Fig. 4.4b:

$$\Omega_1 = \frac{C}{\sqrt{2}m_W}\bar{u}(p_3)\bar{\gamma}^{\nu}\epsilon_{\nu}^{*}(k)u(p_1) \tag{4.33}$$

$$\Omega_2 = \frac{C}{\sqrt{2}m_W}\bar{u}(p_4)\bar{\gamma}^{\mu}\epsilon_{\mu}(k)v(p_2). \tag{4.34}$$

$$\begin{aligned}
\mathcal{M} &= \Omega_1\Omega_2\left(\frac{1}{E_\mu(p_1) - E_{W^-}(q) - E_{\nu_\mu}(p_3)} - \frac{1}{E_{W^+}(-q) + E_{\bar{\nu}_e}(p_2) + E_e(p_4)}\right) \\
&= \Omega_1\Omega_2\left(\frac{1}{E_\mu(p_1) - E_{W^-}(q) - E_{\nu_\mu}(p_3)} - \frac{1}{E_{W^+}(-k) + E_\mu(p_1) - E_{\nu_\mu}(p_3)}\right) \\
&= \frac{C^2}{4m_W^2}\bar{u}(p_4)\gamma^\mu(1-\gamma^5)v(p_2)\,\bar{u}(p_3)\gamma_\mu(1-\gamma^5)u(p_1).
\end{aligned}$$

where last equality follows after polarization sum (Fig. 4.3).

With k as momentum of ν_e and electron momentum as $k + \frac{l}{2}$ as in Fig. 4.4.

Writing

$$\frac{(\hbar c)^6}{V^2} = \frac{k^2\Delta k}{(2\pi)^3}\frac{l^2\Delta l}{(2\pi)^3}d\Omega_1 d\Omega_2 \tag{4.35}$$

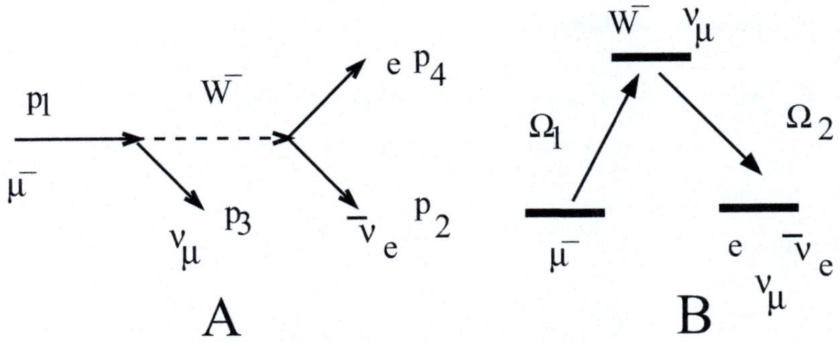

Fig. 4.3 Figure depicts a muon decay $\mu \to \nu_\mu + e + \bar{\nu}_e$

Fig. 4.4 Figure shows momentum conservation in muon decay

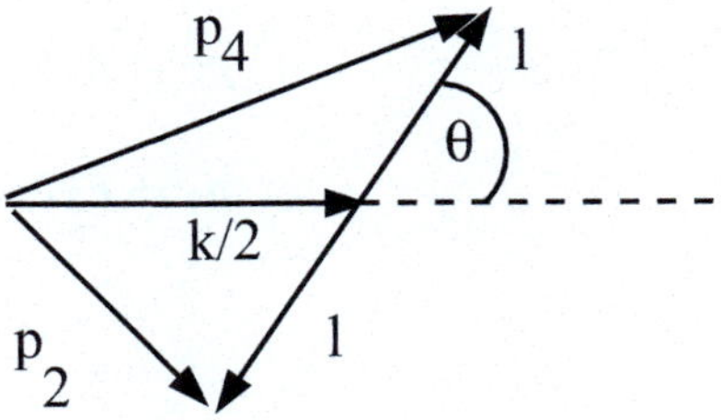

Integrating above over Ω_1 and Ω_2, we get $\frac{1}{8\pi^4}\underbrace{(kl)^2\Delta k\Delta l}_{\Sigma}d\theta$. Let $r=\sqrt{k^2+l^2}$ and $\frac{l}{k}=\tan\theta_1$

$$\Sigma(\theta_1)=r^5\cos^2\theta_1\sin^2\theta_1\Delta r\Delta\theta_1 \tag{4.36}$$

$$\begin{aligned} m_\mu=E=E_2+E_3+E_4=k+\sqrt{\frac{k^2}{4}+l^2+kl\cos\theta} \\ +\sqrt{\frac{k^2}{4}+l^2-kl\cos\theta}=rf(\theta,\theta_1). \end{aligned} \tag{4.37}$$

Decay rate $\Gamma=$

$$\pi\frac{\int\sum_s|\mathcal{M}|^2}{\Delta E}=A\,\alpha_w^2\,\frac{m_\mu^5}{m_W^4}. \tag{4.38}$$

$$A=\frac{2}{\pi}\int_0^\pi\int_0^{\frac{\pi}{2}}\frac{\cos^2\theta_1\sin^2\theta_1}{f^6(\theta,\theta_1)}d\theta_1 d\theta.$$

With $A\sim .01$, $\alpha_w^2\sim10^{-3}$, and $m_w\sim100$ GeV and $m_\mu\sim100$ MeV, we have decay time $\sim\mu$s.

If we care spinor contribution

$$\begin{aligned} \textstyle\sum_s|\mathcal{M}|^2=(\frac{C^2}{16m_W^2})^2\frac{1}{E_1E_2E_3E_4}Tr(\not p_3\gamma^\mu(1-\gamma^5)(\not p_1+m_\mu)\gamma^\nu(1-\gamma^5)) \\ \times Tr((\not p_4+m_e)\gamma_\mu(1-\gamma^5)\not p_2\gamma_\nu(1-\gamma^5)). \end{aligned} \tag{4.39}$$

With a lot of algebra

$$\sum_s|\mathcal{M}|^2=(\frac{C^2}{m_W^2})^2\frac{(p_1\cdot p_2)\,(p_3\cdot p_4)}{E_1E_2E_3E_4} \tag{4.40}$$

$$p_1\cdot p_2=E_1E_2 \tag{4.41}$$

$$p_3\cdot p_4=E_3E_4(1+\frac{l\cos\theta+\frac{k}{2}}{\sqrt{\frac{k^2}{4}+l^2+kl\cos\theta}}) \tag{4.42}$$

$$\frac{(p_1 \cdot p_2)\,(p_3 \cdot p_4)}{E_1 E_2 E_3 E_4} = (1 + \frac{\tan\theta_1 \cos\theta + \frac{1}{2}}{\sqrt{\frac{1}{4} + \tan^2\theta_1 + \tan\theta_1 \cos\theta}}) = g(\theta, \theta_1). \qquad (4.43)$$

Decay rate $\Gamma =$

$$\pi \frac{\int \sum_s |\mathcal{M}|^2}{\Delta E} = A\,\alpha_w^2\,\frac{m_\mu^5}{m_W^4}. \qquad (4.44)$$

$$A = \frac{2}{\pi}\int_0^{\pi}\int_0^{\frac{\pi}{2}} \frac{\cos^2\theta_1 \sin^2\theta_1 g(\theta, \theta_1)}{f^6(\theta, \theta_1)} d\theta_1 d\theta$$

4.5 Pion Decay

Consider charged Pion decay as shown in Fig. 4.5:

$$\pi^- \rightarrow \pi_0 + e + \bar{\nu}_e \qquad (4.45)$$

It is same a muon decay except now instead of emitting a muon neutrino we emit a neutral pion. However, the amplitude of the process is same as in Eq. 4.40:

$$\sum_s |\mathcal{M}|^2 = (\frac{C^2}{m_W^2})^2 (spinor) \qquad (4.46)$$

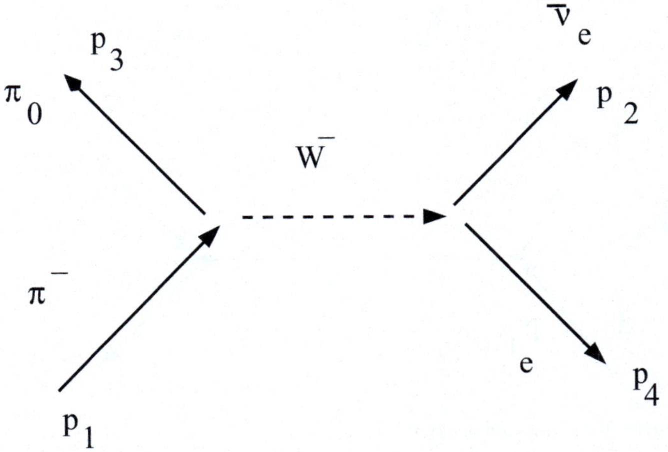

Fig. 4.5 Figure shows pion decay $\pi^- \rightarrow \pi_0 + e + \bar{\nu}_e$

Writing

$$\frac{(\hbar c)^6}{V^2} = \frac{k^2 \Delta k}{(2\pi)^3} \frac{l^2 \Delta l}{(2\pi)^3} d\Omega_1 d\Omega_2 \tag{4.47}$$

$$\Sigma(\theta_1) = r^5 \cos^2\theta_1 \sin^2\theta_1 \Delta r \Delta\theta_1 \tag{4.48}$$

$$\begin{aligned} m_{\pi^-} = E = E_2 + E_3 + E_4 = m_{\pi_0} + \sqrt{\tfrac{k^2}{4} + l^2 + kl\cos\theta} + \sqrt{\tfrac{k^2}{4} + l^2 - kl\cos\theta} \\ = m_{\pi_0} + r f(\theta, \theta_1). \end{aligned} \tag{4.49}$$

Decay rate $\Gamma =$

$$\pi \frac{\int \sum_s |\mathcal{M}|^2}{\Delta E} = A\,\alpha_w^2\, \frac{(m_{\pi^-} - m_{\pi_0})^5}{m_W^4}. \tag{4.50}$$

$$A = \frac{2}{\pi} \int_0^{\pi} \int_0^{\frac{\pi}{2}} \frac{\cos^2\theta_1 \sin^2\theta_1}{f^6(\theta, \theta_1)} d\theta_1 d\theta$$

$A \sim .01$, $\alpha_w^2 \sim 10^{-3}$ and $m_w \sim 100$ GeV and $m_{\pi^-} \sim 139$ MeV, $m_{\pi_0} \sim 135$ MeV, we have decay time ~s (Fig. 4.6).

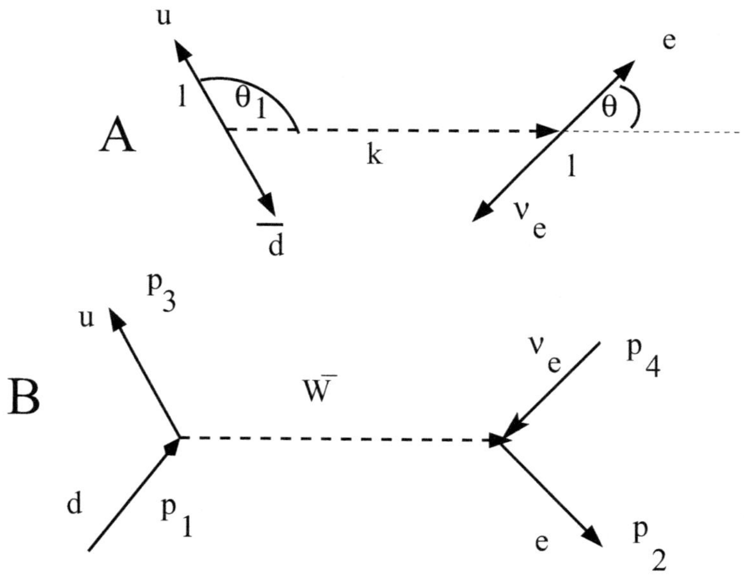

Fig. 4.6 Figure shows the pion decay $\pi^- \to e + \bar{\nu}_e$

If we care the spinor part

$$\sum_s |\mathcal{M}|^2 = (\frac{C^2}{m_W^2})^2 \frac{(p_1 \cdot p_2)\,(p_3 \cdot p_4)}{E_1 E_2 E_3 E_4} \tag{4.51}$$

$$p_1 \cdot p_2 = E_1 E_2 \tag{4.52}$$

$$p_3 \cdot p_4 = E_3 E_4 (1 + k \frac{l \cos\theta + \frac{k}{2}}{\sqrt{m_e^2 + \frac{k^2}{4} + l^2 + kl\cos\theta}\sqrt{m_{\pi_0}^2 + k^2}}) \tag{4.53}$$

$$\sim E_3 E_4 (1 + \frac{k}{m_{\pi_0}} \frac{l \cos\theta + \frac{k}{2}}{\sqrt{\frac{k^2}{4} + l^2 + kl\cos\theta}}) \tag{4.54}$$

$$\frac{(p_1 \cdot p_2)\,(p_3 \cdot p_4)}{E_1 E_2 E_3 E_4} = (1 + \frac{r\cos\theta_1}{m_{\pi_0}} \frac{\tan\theta_1 \cos\theta + \frac{1}{2}}{\sqrt{\frac{1}{4} + \tan^2\theta_1 + \tan\theta_1\cos\theta}})$$

$$= (1 + \frac{r}{m_{\pi_0}} g(\theta, \theta_1)). \tag{4.55}$$

$$\pi \frac{\int \sum_s |\mathcal{M}|^2}{\Delta E} = A\,\alpha_w^2\, \frac{(m_{\pi^-} - m_{\pi_0})^5}{m_W^4}. \tag{4.56}$$

$$A = \frac{2}{\pi} \int_0^{\pi} \int_0^{\frac{\pi}{2}} \cos^2\theta_1 \sin^2\theta_1 \frac{1 + \frac{\frac{m_{\pi^-}}{m_{\pi_0}} - 1}{f} g(\theta, \theta_1)}{f^5(\theta, \theta_1)} d\theta_1 d\theta$$

4.6 More Pion Decay

$$\pi^- \to e + \bar{\nu}_e \tag{4.57}$$

$$d + \bar{u} \xrightarrow{W^-} e + \bar{\nu}_e \tag{4.58}$$

π^- is a bound state of d quark and u antiquark. The bound state can be written as sum of states like (Fig. 4.7)

$$\phi = \exp(ik \cdot (\frac{r_1 + r_2}{2}))\exp(il \cdot (\frac{r_1 - r_2}{2})) = \exp(ip_1 r_1)\exp(-ip_3 r_2)$$

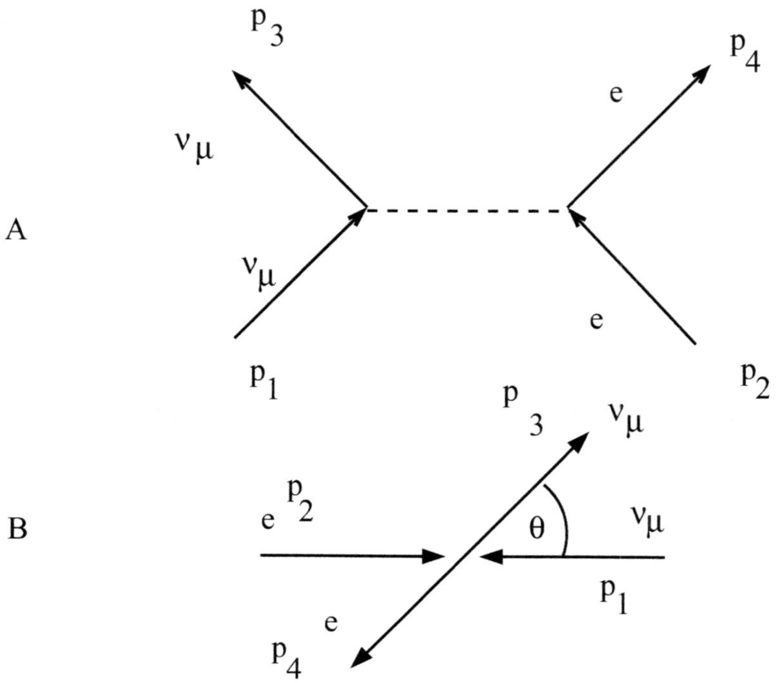

Fig. 4.7 Figure shows the neutral weak scattering $\nu_e + \mu \to \nu_e + \mu$

with different $l's$ as shown in Fig. 4.8a corresponding to different θ_1. Then, $p_1 = k/2 + l$ and $-p_3 = k/2 - l$ as in Fig. 4.8b. For pion at rest, $k = 0$. The energy of the pion then is

$$\begin{aligned} m_\pi &= E = \sqrt{m_u^2 + p_1^2} + \sqrt{m_d^2 + p_3^2} \\ &= \sqrt{m_u^2 + \tfrac{k^2}{4} + l^2 + kl\cos\theta_1} + \sqrt{m_d^2 + \tfrac{k^2}{4} + l^2 - kl\cos\theta_1} \sim 2l \end{aligned} \tag{4.59}$$

when $k = 0$.

Let us calculate the decay rate for one configuration $\theta_1 = 0$. The total decay rate then is the average over θ_1 which by symmetry is just as for $\theta_1 = 0$.

$$\mathcal{M} \sim \frac{C^2}{m_W^2}(spinor) \tag{4.60}$$

$$m_\pi = E = E_3 + E_4 = \sqrt{m_e^2 + \frac{k^2}{4} + l^2 + kl\cos\theta} + \sqrt{\frac{k^2}{4} + l^2 - kl\cos\theta} \sim 2l. \tag{4.61}$$

$$\Delta E = 2\Delta l \tag{4.62}$$

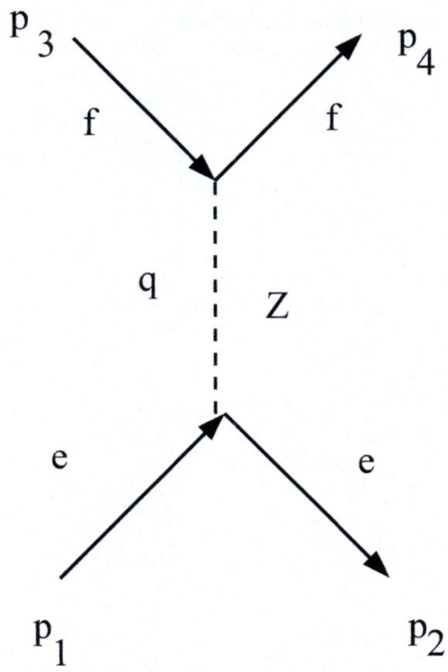

Fig. 4.8 Figure shows weak electron-positron scattering

$$\frac{(\hbar c)^3}{V^3} = \frac{l^2 \Delta l}{(2\pi)^3} d\Omega_1 \tag{4.63}$$

Decay rate $\Gamma =$

$$\pi \frac{\int \sum_s |\mathcal{M}|^2}{\Delta E} = \frac{\pi}{2} \frac{\alpha_w^2 m_\pi^2}{m_W^4} \frac{(\hbar c)^3}{V_0} \tag{4.64}$$

With V_0 corresponding to pion radius of 1 fm, we find $\frac{(\hbar c)^3}{V_0} \sim 200(Mev)^3$ and decay time $\sim 10^{-8}$s.

If we care spinors

$$\mathcal{M} \sim \frac{C^2}{m_W^2} \bar{u}(p_4)\bar{\gamma}^\mu u(p_2)\, \bar{v}(p_3)\bar{\gamma}_\mu u(p_1) \tag{4.65}$$

$$= \frac{C^2}{4m_W^2} \bar{u}(p_4)\gamma^\mu(1-\gamma^5)u(p_2)\, \bar{v}(p_3)\gamma_\mu(1-\gamma^5)u(p_1). \tag{4.66}$$

$$\sum_s |\mathcal{M}|^2 = (\frac{C^2}{16m_W^2})^2 \frac{1}{E_1E_2E_3E_4} Tr(\not{p}_3\gamma^\mu(1-\gamma^5)(\not{p}_1 + m_e)\gamma^\nu(1-\gamma^5)) \\ \times Tr((\not{p}_4 + m_\mu)\gamma_\mu(1-\gamma^5)\not{p}_2\gamma_\nu(1-\gamma^5)). \tag{4.67}$$

With a lot of algebra

$$\sum_s |\mathcal{M}|^2 = (\frac{C^2}{m_W^2})^2 \frac{(p_1 \cdot p_2)\,(p_3 \cdot p_4)}{E_1 E_2 E_3 E_4} \tag{4.68}$$

$$= (\frac{C^2}{m_W^2})^2 (1 - \cos^2\theta) \tag{4.69}$$

$$= (\frac{C^2}{m_W^2})^2 \sin^2\theta \tag{4.70}$$

$$\pi \frac{\int \sum_s |\mathcal{M}|^2}{\Delta E} = \frac{\pi}{2} \frac{\alpha_w^2 m_\pi^2}{m_W^4} \frac{(\hbar c)^3}{V_0} \int \sin^2\theta d\theta. \tag{4.71}$$

$$= (\frac{\pi}{2})^2 \frac{(\hbar c)^3}{V_0} \frac{\alpha_w^2 m_\pi^2}{m_W^4} \tag{4.72}$$

4.7 Neutral Weak Interactions

4.7.1 Elastic Neutrino-Electron Scattering

$$\nu_\mu + e \xrightarrow{Z} \nu_\mu + e \tag{4.73}$$

This is mediated by a Z boson, as no charge exchange takes place in interaction:

$$\mathcal{M} \sim \frac{C^2}{m_Z^2} (spinor) \tag{4.74}$$

We can just write the cross section as $l = \frac{\alpha_w \hbar c E}{m_Z^2}$, which at GeV energy is 10 femto-barn.

If we care spinor, the Z vertex is bit more complicated as we will see subsequently, we can say, it is combination of left spinor and right spinor:

$$\mathcal{M} \sim \frac{C^2}{4m_Z^2} \bar{u}(p_4)\gamma^\mu (c_V - c_A\gamma^5) u(p_2)\, \bar{u}(p_3)\gamma_\mu (1 - \gamma^5) u(p_1). \tag{4.75}$$

$$\sum_s |\mathcal{M}|^2 = (\frac{C^2}{16 m_Z^2})^2 \frac{1}{E_1 E_2 E_3 E_4}$$
$$\times Tr((\not{p}_3 + m_e)\gamma^\mu (c_V - c_A\gamma^5)(\not{p}_1 + m_e)\gamma^\nu (c_V - c_A\gamma^5))$$
$$\times\ Tr(\not{p}_4 \gamma_\mu (1 - \gamma^5)\not{p}_2 \gamma_\nu (1 - \gamma^5)).$$

With a lot of algebra and E as CM energy

$$\sum_s |\mathcal{M}|^2$$

$$= (\frac{C^2}{2m_Z^2})^2 \frac{((c_A + c_V)^2 (p_1 \cdot p_2)\,(p_3 \cdot p_4) + ((c_A - c_V)^2 (p_1 \cdot p_4)\,(p_3 \cdot p_3) + m_e^2 (c_A^2 - c_V^2)(p_1 \cdot p_3)}{E_1 E_2 E_3 E_4}$$

$$= (\frac{C^2}{m_Z^2})^2 ((c_A + c_V)^2 + ((c_A - c_V)^2 \cos^4 \frac{\theta}{2})$$

$$\frac{d\sigma}{d\theta} = 4\pi (\frac{\alpha_w \hbar c E}{m_Z^2})^2 ((c_A + c_V)^2 + ((c_A - c_V)^2 \cos^4 \frac{\theta}{2})$$

4.7.2 Electron Positron Scattering

Boson-mediated interaction has amplitude:

$$\mathcal{M} = \frac{C^2}{4(q^2 - m_Z^2)} (spinor). \tag{4.76}$$

We can just write the cross section as

$$l = \frac{\alpha_w \hbar c E}{M_z^2},$$

which is as shown before, 10 femto-barn at energy 1 GeV.

If we care all spinors

$$\mathcal{M} = \frac{C^2}{4(q^2 - m_Z^2)} \bar{u}(p_4) \gamma^\mu (c_V^f - c_A^f \gamma^5) v(p_3)\; \bar{v}(p_2) \gamma_\mu (c_V^e - c_A^e \gamma^5) u(p_1). \tag{4.77}$$

$$\sum_s |\mathcal{M}|^2 = (\frac{C^2}{16(q^2 - m_Z^2)})^2 \frac{1}{E_1 E_2 E_3 E_4} Tr(\not{p}_4 \gamma^\mu (c_V^f - c_A^f \gamma^5) \not{p}_1 \gamma^\nu (c_V^f - c_A^f \gamma^5))$$
$$\times\, Tr(\not{p}_2 \gamma_\mu (c_V^e - c_A^e \gamma^5) \not{p}_1 \gamma_\nu (c_V^e - c_A^e \gamma^5)).$$

$$\sum_s |\mathcal{M}|^2 = \frac{1}{2} (\frac{C^2}{2(q^2 - m_Z^2)})^2 \frac{1}{E_1 E_2 E_3 E_4} \{\, (\, (c_A^e)^2 + (c_V^e)^2\,)(\, (c_A^f)^2 + (c_V^f)^2)$$
$$[(p_1 \cdot p_2)\,(p_3 \cdot p_4) + (p_1 \cdot p_4)\,(p_2 \cdot p_3)]$$
$$+4 c_V^e c_A^e c_V^f c_A^f\, [(p_1 \cdot p_2)\,(p_3 \cdot p_4) - (p_1 \cdot p_4)\,(p_3 \cdot p_3)]\,\}$$

In CM frame, it reduces to

$$\sum_s |\mathcal{M}|^2 = (\frac{C^2}{2((2E)^2 - m_Z^2)})^2 \frac{1}{E_1 E_2 E_3 E_4} \{ ((c_A^e)^2 + (c_V^e)^2)((c_A^f)^2 + (c_V^f)^2)$$

$$\left[(1 + \cos^2\theta)\right]$$

$$-8 c_V^e c_A^e c_V^f c_A^f \cos\theta \}$$

The differential cross section

$$\sum_s |\mathcal{M}|^2 = (\frac{C^2}{2((2E)^2 - m_Z^2)})^2 \{((c_A^e)^2 + (c_V^e)^2)((c_A^f)^2 + (c_V^f)^2) \left[(1 + \cos^2\theta)\right]$$
$$- 8 c_V^e c_A^e c_V^f c_A^f \cos\theta \}$$

The differential cross section is

$$\frac{d\sigma}{d\theta} = \pi(\frac{\alpha_w \hbar c E}{(2E)^2 - m_Z^2})^2 \{((c_A^e)^2 + (c_V^e)^2)((c_A^f)^2 + (c_V^f)^2) \left[(1 + \cos^2\theta)\right]$$
$$- 8 c_V^e c_A^e c_V^f c_A^f \cos\theta \}$$

4.8 Electroweak Unification, Parity Violation, and Mass

4.8.1 Introduction

Beginning with the seminal work of Yang and Lee [1] and its experimental verification by Wu [2], it is well known that weak interactions do not preserve parity. In the theory of weak interactions, this is manifested by coupling only the left-handed components of the fermion doublet. The work of parity violation began with Yang and Lee's observations on K-mesons which led them to question parity conservation in weak interactions. This led them to devise many experiments that would test parity conservation. The first of these was carried out by Wu [2], which confirmed parity violation in weak interactions.

The basic experiment of Wu was β decay of a cobalt CO^{60} nucleus that had its nuclear spin oriented by magnetic field along z direction. After β decay, the nucleus changed to Ni^{60} by neutron changing to proton. Electron and neutrino were emitted, with both having spin along z direction, but electron (relativistic) could have been moving along or opposite to z, and it always turned out it was along z , which is violation of parity. The explanation now is of course obvious the vertex has projection on the left spinor so we do not see the alternative. Figure 4.9 shows the cobalt decay experiment of Wu, where electrons are always emitted in one direction.

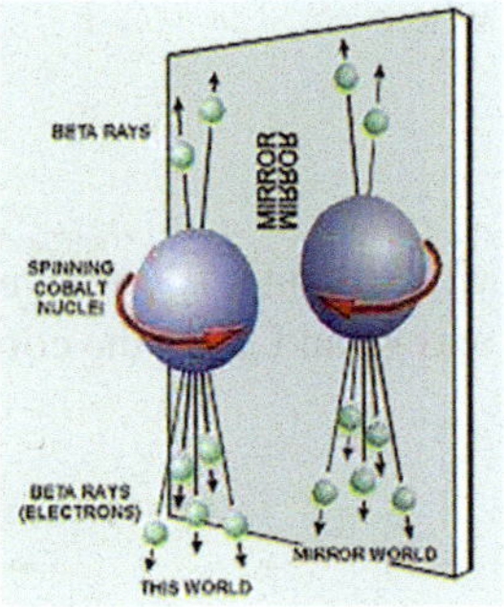

Fig. 4.9 Figure shows the cobalt decay experiment of Wu, where electrons are always emitted in one direction

Further developments in the theory of weak interactions include invention of Higgs mechanism which gives masses to vector bosons and fermions [3–5] and the theory of electroweak unification [6, 7]. Historical facts suggest that work on parity violation preceded the work on Higgs mechanism and electroweak unification. In this chapter, we take a different viewpoint. We suggest that parity violation in weak interactions can be predicted on pure theoretical grounds. In this paper, we show that parity violation is a natural consequence of gauge invariance. In a theory where there is no parity violation, we cannot assign masses to fermions in a gauge-invariant way using the Higgs mechanism because Higgs field transforms in a quadratic way under gauge transformation. However, when we violate parity and only couple the left-handed components of the fermions, Higgs field transforms in a linear way under gauge transformation, and it becomes possible to give masses to fermions in a gauge-invariant manner.

The section is organized as follows. We first review the basics of Higgs mechanism for giving masses to vector bosons and fermions [8–11]. We then go through the exercise of showing how the theory is gauge invariant, when we have parity violation. Then, we work through a theory where there is no parity violation and show we cannot assign masses to fermions in a gauge-invariant way using the Higgs mechanism.

4.8.2 *Theory*

We consider the Higgs doublet:

$$\Phi = \begin{bmatrix} \Phi_A \\ \Phi_B \end{bmatrix}. \tag{4.78}$$

The field is coupled to electromagnetic field and W, Z bosons with gauge coupling, with Lagrangian density:

$$\mathcal{L}_\Phi = D_\mu \Phi^\dagger D^\mu \Phi - V(\Phi^\dagger \Phi), \tag{4.79}$$

where

$$D_\mu = \partial_\mu + i\frac{g_1}{2}B_\mu + i\frac{g_2}{2}\mathbf{W}_\mu, \tag{4.80}$$

with B_u and $\mathbf{W}_\mu$ the vector potential for EM and weak interactions, respectively, and g_1 and g_2 as the corresponding coupling constants and

$$V(\Phi^\dagger\Phi) = \frac{m^2}{2\phi_0^2}[(\Phi^\dagger\Phi) - \phi_0^2]^2, \tag{4.81}$$

where the ground state of the Higgs field is

$$\Phi_{ground} = \begin{bmatrix} 0 \\ \phi_0 \end{bmatrix}, \tag{4.82}$$

and the excited state

$$\Phi = \begin{bmatrix} 0 \\ \phi_0 + \frac{h(x)}{\sqrt{2}} \end{bmatrix}. \tag{4.83}$$

Substituting Φ in Eq. 4.79 gives masses to W, Z bosons and the Higgs boson via gauge coupling:

$$D_\mu\Phi = \begin{pmatrix} 0 \\ \frac{\partial_\mu h}{\sqrt{2}} \end{pmatrix} + i\frac{g_1}{2}\begin{pmatrix} 0 \\ B_\mu(\phi_0 + \frac{h(x)}{\sqrt{2}}) \end{pmatrix} + i\frac{g_2}{2}\begin{pmatrix} \sqrt{2}W_\mu^+(\phi_0 + \frac{h(x)}{\sqrt{2}}) \\ -W_\mu^3(\phi_0 + \frac{h(x)}{\sqrt{2}}) \end{pmatrix}, \tag{4.84}$$

and

$$\begin{aligned}\mathcal{L}_\Phi &= \tfrac{1}{2}\partial_\mu h\partial^\mu h + \tfrac{g_2^2}{4}(W_\mu^{+\prime}W^{+\mu} + W_\mu^{-\prime}W^{-\mu})(\phi_0 + \tfrac{h(x)}{\sqrt{2}})^2 \\ &+ \tfrac{g_1^2+g_2^2}{4}Z_\mu Z^\mu(\phi_0 + \tfrac{h(x)}{\sqrt{2}})^2 - V(h),\end{aligned} \tag{4.85}$$

where $W_\mu^+ = \frac{W_\mu^1 - iW_\mu^2}{\sqrt{2}}$ and $W_\mu^- = \frac{W_\mu^1 + iW_\mu^2}{\sqrt{2}}$ are the W bosons and

$$Z_\mu = W_\mu^3\cos\theta_w - B_\mu\sin\theta_w, \tag{4.86}$$

the Z boson and the massless photon

$$A_\mu = W_\mu^3\sin\theta_w + B_\mu\cos\theta_w, \tag{4.87}$$

where θ_w is the Weinberg angle

$$\cos\theta_w = \frac{g_2}{\sqrt{g_1^2+g_2^2}}, \quad \sin\theta_w = \frac{g_1}{\sqrt{g_1^2+g_2^2}}. \tag{4.88}$$

The field couples to fermions as follows. Let us consider the neutrino-electron doublet written as a four vector:

$$L = \begin{bmatrix} \nu_R \\ \nu_L \\ e_L \\ e_R \end{bmatrix}. \tag{4.89}$$

Using the notation $\sigma_j = (\sigma_x, \sigma_y, \sigma_z)$, $\sigma_u = (\sigma_0, \sigma_x, \sigma_y, \sigma_z)$, and $\tilde{\sigma}_u = (\sigma_0, -\sigma_x, -\sigma_y, -\sigma_z)$, the doublet evolves ($\hbar$ and c are implicit) as $i\frac{dL}{dt} =$

$$\begin{bmatrix} i\partial_j\sigma_j & m_\nu & 0 & 0 \\ m_\nu & -i\partial_j\sigma_j + \frac{1}{2}(g_2 W_\mu^3 - g_1 B_\mu)\sigma_\mu & \frac{g_2}{\sqrt{2}} W_\mu^+\sigma_\mu & 0 \\ 0 & \frac{g_2}{\sqrt{2}} W_\mu^-\sigma_\mu & -i\partial_j\sigma_j - \frac{1}{2}(g_2 W_\mu^3 + g_1 B_\mu)\sigma_\mu & m_e \\ 0 & 0 & m_e & i\partial_j\sigma_j - g_1 B_\mu\tilde{\sigma}_\mu \end{bmatrix} \begin{bmatrix} \nu_R \\ \nu_L \\ e_L \\ e_R \end{bmatrix}. \tag{4.90}$$

where $m_e = c_e(\phi_0 + \frac{h(x)}{\sqrt{2}})$ and $m_\nu = c_\nu(\phi_0 + \frac{h(x)}{\sqrt{2}})$, with c_e, c_ν as coupling of electron and neutrino to Higgs boson.

When we express the above equation in terms of the fields Z_μ, A_μ, it takes the form:
$i\frac{dL}{dt} =$

$$\begin{bmatrix} i\partial_j\sigma_j & m_\nu & 0 & 0 \\ m_\nu & -i\partial_j\sigma_j + \frac{e}{\sin 2\theta_w} Z_\mu\sigma_\mu & \frac{g_2}{\sqrt{2}} W_\mu^+\sigma_\mu & 0 \\ 0 & \frac{g_2}{\sqrt{2}} W_\mu^-\sigma_\mu & -i\partial_j\sigma_j - e(A_\mu + \cot 2\theta_w Z_\mu)\sigma_\mu & m_e \\ 0 & 0 & m_e & i\partial_j\sigma_j - e(A_\mu - \tan\theta_w Z_\mu)\tilde{\sigma}_\mu \end{bmatrix} \begin{bmatrix} \nu_R \\ \nu_L \\ e_L \\ e_R \end{bmatrix}. \tag{4.91}$$

where $g_1\cos\theta_w = g_2\sin\theta_w = e$, with $-e$, the electron charge.

Now, we look at how equations 4.79 and 4.90 transform when we make a gauge transformation on $\mathbf{W}$ and B. The transformations are for $U \in SU(2)$, we have

$$\mathbf{W}_\mu \to U(x)\mathbf{W}_\mu U^\dagger(x) + \frac{i\partial_\mu U(x)U^\dagger(x)}{g_2/2}, \tag{4.92}$$

$$B_\mu \to B_\mu - \frac{\partial_\mu \theta(x)}{g_1/2}. \tag{4.93}$$

Then, the Higgs doublet transforms as

$$\Phi \to \Theta(x)\Phi \tag{4.94}$$

where $\Theta(x) = \exp(i\theta(x))U(x)$.

In terms of field Φ, the equation for L takes the form $i\frac{dL}{dt} =$

$$\begin{bmatrix} i\partial_j\sigma_j & c_\nu\Phi_A^* & c_\nu\Phi_B^* & 0 \\ c_\nu\Phi_A & -i\partial_j\sigma_j + \frac{1}{2}(g_2W_\mu^3 - g_1B_\mu)\sigma_\mu & \frac{g_2}{\sqrt{2}}W_\mu^+\sigma_\mu & -c_e\Phi_B^* \\ c_\nu\Phi_B^* & \frac{g_2}{\sqrt{2}}W_\mu^-\sigma_\mu & -i\partial_j\sigma_j - \frac{1}{2}(g_2W_\mu^3 + g_1B_\mu)\sigma_\mu & c_e\Phi_A^* \\ 0 & -c_e\Phi_B & c_e\Phi_A & i\partial_j\sigma_j - g_1B_\mu\tilde{\sigma}_\mu \end{bmatrix}$$
$$\begin{bmatrix} \nu_R \\ \nu_L \\ e_L \\ e_R \end{bmatrix}. \tag{4.95}$$

where under the gauge transformation L transforms as

$$\begin{bmatrix} \nu_L \\ e_L \end{bmatrix} \to \exp(-i\theta(x))U(x)\begin{bmatrix} \nu_L \\ e_L \end{bmatrix} \tag{4.96}$$

$$e_R \to \exp(-i2\theta(x))e_R \tag{4.97}$$

In Eq. 4.90, only e_L and ν_L are coupled. e_R and ν_R are not coupled. That is to say, we have parity violation. We now show that this physical law is in fact a consequence of the fact that it is not possible to give masses to fermions in a manner that is gauge-invariant (as above), if we do not violate parity.

To see this, let's reorganize the doublet as

$$M = \begin{bmatrix} \nu \\ e \end{bmatrix}, \quad \nu = \begin{bmatrix} \nu_L \\ \nu_R \end{bmatrix}, \quad e = \begin{bmatrix} e_L \\ e_R \end{bmatrix}. \tag{4.98}$$

$$i\frac{dM}{dt} = \left(\underbrace{\begin{bmatrix} -i\partial_j\alpha_j + \frac{1}{2}(g_2 W_\mu^3 - g_1 B_\mu)\alpha_\mu & \frac{g_2}{\sqrt{2}} W_\mu^+ \alpha_\mu \\ \frac{g_2}{\sqrt{2}} W_\mu^- \alpha_\mu & -i\partial_j\alpha_j - \frac{1}{2}(g_2 W_\mu^3 + g_1 B_\mu)\alpha_\mu \end{bmatrix}} + \underbrace{\begin{bmatrix} m_\nu & 0 \\ 0 & m_e \end{bmatrix}}_{C} \beta \right) \begin{bmatrix} \nu \\ e \end{bmatrix}, \tag{4.99}$$

where $\beta = \sigma_x \otimes \sigma_0$ and $\alpha_u = \sigma_z \otimes \sigma_\mu$ (α_0 is identity), with σ_μ Pauli matrices. If we plan to write this Eq. 4.99, in terms of Higgs field Φ, then we find that Φ enters the term C above. To make it gauge invariant, this term should be of the form:

$$C(\Phi) = \Theta(x) \begin{bmatrix} m_\nu & 0 \\ 0 & m_e \end{bmatrix} \Theta^\dagger(x), \tag{4.100}$$

In the above, $C(\Phi)$ cannot be expressed in terms of Φ alone. The best we can write it is

$$C(\Phi) = \begin{bmatrix} c_\nu \Phi_B^* & c_e \Phi_A \\ -c_\nu \Phi_A^* & c_e \Phi_B \end{bmatrix} \begin{bmatrix} \exp(i\theta(x)) & 0 \\ 0 & \exp(-i\theta(x)) \end{bmatrix} U^\dagger(x), \tag{4.101}$$

which is still not just Φ dependent. Hence, when $m_e \neq m_\nu$, we cannot make our equations gauge invariant unless we do a *parity violation*. Therefore, parity violation arises as a consequence of gauge invariance.

4.9 Gauge Potential

Energy of gauge potential W_μ is

$$W_{\mu\nu} = F_{\mu\nu} + \frac{i}{g}[W_\mu, W_\nu]$$

$$F_{\mu\nu} = \partial_\mu W_\nu - \partial_\nu W_\mu$$

$$E = tr\ W_{\mu,\nu} W^{\mu,\nu}$$

If we define gauge transformation as

$$W \to W' = UWU' - ig\ \partial\mu U\ U^\dagger.$$

Then, we have $W'_{\mu,\nu} = U W_{\mu,\nu} U^\dagger$ and energy doesnot take.

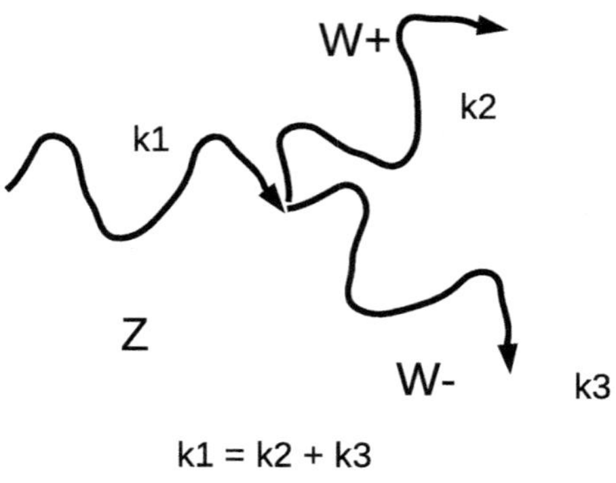

Fig. 4.10 Figure shows a 3 vertex, a Z photon transit to $W+$ and $W-$ photon, the interaction energy is as in Eq. 4.102

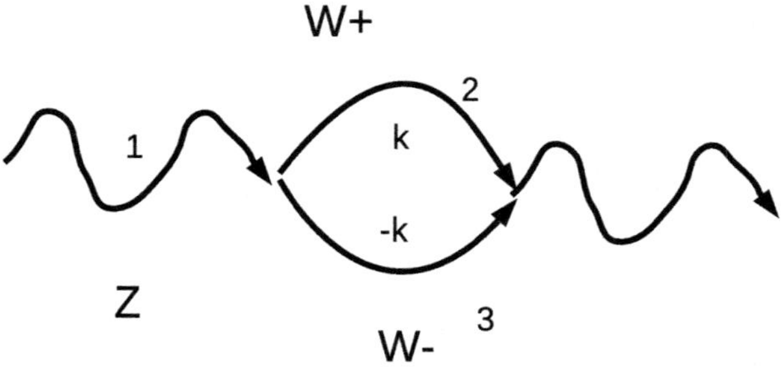

Fig. 4.11 Figure shows a Z photon transit to $W+$ and $W-$ photon, which recombine to give back Z boson

Now, if we have three photons 1, 2, 3, whose gauge potential forms a $su(2)$ algebra, then the interaction energy density of the three photon can be written as $E = E_1 + E_2$:

$$E = \frac{1}{g} W_3^{\dagger} F_{\mu\nu}^1 W_2. \tag{4.102}$$

If the momentum of three photons adds to zero, then density integrates to finite interaction energy (Fig. 4.10).

4.10 Renormalizing the W-Z Boson Mass

Figure 4.11 shows a Z photon transit to $W+$ and $W-$ photon, which recombine to give back Z boson. There is correction to the boson energy due to this Feynman diagram.

We calculate this if limit $|k|$ becomes large, then in expression for E scale like $\frac{1}{E_k}$, then the total amplitude

$$\mathcal{M} = \eta \frac{E_2^2}{E_k} = \frac{\eta}{E_k^3} \sim \frac{1}{E_K^4},$$

which is summable and renormalizable.

The effective mass of boson increases as result of this normalization:

$$M_Z' > M_z.$$

4.11 Problems

1. Mass of W boson is $M_W = 80$ GeV/c^2. Find its energy if its momentum $p = 1$ GeV/c.
2. Mass of Z boson is $M_Z = 90$ GeV/c^2. Find its energy if its wavelength is 1 femto meter.
3. In inverse muon decay, electron and muon neutrino collide at energy of 1 Gev each along z axis. Find the differential cross section of muon production given mass $m_\mu = 107$ MeV/c^2.
4. Find the decay rate of a muon to electron and antineutrino; in particular, find numerically the coefficient A discussed in the chapter.
5. Find the decay rate of

$$\pi^- \to \pi^0 + e + \bar{\nu_e}.$$

In particular, find numerically the coefficient A discussed in the chapter. $m_{\pi^-} = 139$ MeV/c^2. $m_{\pi^0} = 135$ MeV/c^2.

In above, take weak fine structure constant $\alpha_w = \frac{1}{29}$.

References

1. T.D. Lee, C.N. Yang, Question of parity conservation in weak interactions. Phys. Rev., **104**, 254 (1956)
2. C.S. Wu, E. Ambler, R.W. Hayward, D.D. Hoppes, R.P. Hudson, Experimental test of parity conservation in beta decay. Phys. Rev. **105**, 1413 (1957)
3. H. Peter, Broken symmetries and the masses of gauge bosons. Phys. Rev. Lett. **13**, (1964) (16)
4. F. Englert, R. Brout, Broken symmetry and the mass of gauge vector mesons. Phys. Rev. Lett. **13**(9), 321–23 (1964)
5. P.W. Anderson, Plasmons, gauge invariance, and mass. Phys. Rev. **130**, 439 (1963)
6. A. Salam, J.C. Ward, Weak and electromagnetic interactions. Nuovo Cimento **11**(4), 568–577 (1959)

7. S. Weinberg, A model of leptons. Phys. Rev. Lett. **19**, 1264–66 (1967)
8. W.A. Cottingham, D.A. Greenwood, *An Introduction to Standard Model of Particle Physics* (Cambridge University Press, 2007)
9. M.E. Peskin, D.V. Schroeder, *An Introduction to Quantum Field Theory* (Westview Press, 1995)
10. D. Griffiths, *Introduction to Elementary Particles* (Wiley-VCH, 2004)
11. D.J. Griffiths, *Introduction to Quantum Mechanics* (Pearson Prentice Hall, 2004)

Chapter 5
Quantum Chromodynamics

5.1 Quarks, Color, and Gluons

The nucleus of an atom is made of protons and neutrons. Protons and neutrons are themselves divisible and composed of elementary particle quarks. Quarks can have charge $\frac{2}{3}e$ as in quarks u, c, t or charge $-\frac{1}{3}e$ as in quarks d, s, b. As $electron, neutrino$ forms a doublet $\begin{pmatrix} e \\ \nu \end{pmatrix}$ which interacts through weak force we have the doublets $\begin{pmatrix} u \\ d \end{pmatrix}$, $\begin{pmatrix} c \\ s \end{pmatrix}$ and $\begin{pmatrix} t \\ b \end{pmatrix}$ forms a doublet which interacts through weak force.

Baryons are composed of three quarks, like proton is uud and neutron udd. Similarly, mesons are composed of a quark and antiquark like pion π^+ is $u\bar{d}$ and π^- is $d\bar{u}$, kaon K^+ is $u\bar{s}$, and K^- is $\bar{u}s$.

Quarks also possess like charge another property called color. Quarks come in three color r, g, b or $red, green, blue$; we write the state of a quarks as

$$r = \begin{pmatrix} 1 \\ 0 \\ 0 \end{pmatrix}; \quad b = \begin{pmatrix} 0 \\ 1 \\ 0 \end{pmatrix}; \quad g = \begin{pmatrix} 0 \\ 0 \\ 1 \end{pmatrix} \tag{5.1}$$

The color states are states in three-dimensional space. Transition between them is mediated by six gluons, which can be represented as $SU(3)$ generators , which are $su(3)$ matrices, the eight-dimensional space consisting of matrices. The extra two diagonal generators give two extra gluons making in all 8 gluons.

$$\lambda^1 = \begin{pmatrix} 0 & 1 & 0 \\ 1 & 0 & 0 \\ 0 & 0 & 0 \end{pmatrix}, \quad \lambda^2 = \begin{pmatrix} 0 & -i & 0 \\ i & 0 & 0 \\ 0 & 0 & 0 \end{pmatrix}, \quad \lambda^3 = \frac{1}{\sqrt{2}} \begin{pmatrix} 1 & 0 & 0 \\ 0 & -1 & 0 \\ 0 & 0 & 0 \end{pmatrix}$$

N. Khaneja, *High Energy Physics*, https://doi.org/10.1007/978-3-031-75645-0_5

$$\lambda^4 = \begin{pmatrix} 0 & 0 & 1 \\ 0 & 0 & 0 \\ 1 & 0 & 0 \end{pmatrix}, \quad \lambda^5 = \begin{pmatrix} 0 & 0 & i \\ 0 & 0 & 0 \\ -i & 0 & 0 \end{pmatrix}, \lambda^6 = \begin{pmatrix} 0 & 0 & 0 \\ 0 & 0 & 1 \\ 1 & 0 & 0 \end{pmatrix}$$

$$\lambda^7 = \begin{pmatrix} 0 & 0 & 0 \\ 0 & 0 & -i \\ 0 & i & 0 \end{pmatrix}, \lambda^8 = \frac{1}{\sqrt{6}} \begin{pmatrix} 1 & 0 & 0 \\ 0 & 1 & 0 \\ 0 & 0 & -2 \end{pmatrix}$$

λ_j is what we call gauge potential; the gluon that transits b to g written as $\bar{b}g$ takes the following form which can be written as a gauge potential.

Consider the matrix:

$$\begin{pmatrix} 0 & \exp(i(kx - \omega t)) & 0 \\ \exp(-i(kx - \omega t)) & 0 & 0 \\ 0 & 0 & 0 \end{pmatrix} = \cos(kx - \omega t)\lambda_1 + \sin(kx - \omega t)\lambda_2$$

This corresponds to a gluon $\bar{b}r$ with momentum k that makes a transition from b to r by absorbing the gluon, or we can transit from r to b by emitting this gluon. In the first case, the momentum increases by k and in the second case decreases by k.

Consider the matrix:

$$\cos(kx - \omega t)\lambda_4 + \sin(kx - \omega t)\lambda_5 = \begin{pmatrix} 0 & 0 & \exp(i(kx - \omega t)) \\ 0 & 0 & 0 \\ \exp(-i(kx - \omega t)) & 0 & 0 \end{pmatrix}$$

This corresponds to a quark $\bar{g}r$ with momentum k that makes a transition from g to b by absorbing the gluon, or we can transit from r to g by emitting this gluon. In the first case, the momentum increases by k and in the second case decreases by k.

How does this transition take place. The state of the g quark is a spinor $u_1 \exp(ipx)$ with momentum k, and let $q = p + k$, then we make transition to red spinor $u_2 \exp(iqx)$, with amplitude that is

$$\mathcal{M} = \frac{C}{\sqrt{2E_k}} \bar{u}_2 \epsilon_\mu \gamma^\mu u_1 \tag{5.2}$$

where $C = \frac{\hbar c g_s}{\sqrt{V}}$ with g_s strong coupling constant and ϵ_μ gluon polarization.

5.2 Quark-Quark Interaction and Color Factor

5.2.1 *Quark-Antiquark Interaction*

Let us consider a meson with quark and antiquark pair. Let us hypothesize the color state be say $r\bar{g}$. Figure 5.1 shows the Feynman diagram for quark-antiquark interaction. c_i are the colors on quarks with $c_1 = c_3 = r$ and $c_2 = c_4 = \bar{b}$. Red scatters to red and blue to blue. This can be accounted by two set of gluons as in λ^3 and λ^8. The resulting amplitude is as in electron-electron scattering:

$$\mathcal{M} = C^2 \frac{\bar{u}_3 \gamma^\mu u_1 \bar{v}_4 \gamma_\mu v_2}{q^2} \left(c_1' \lambda^3 c_1 c_2' \lambda^3 c_2 + c_1' \lambda^8 c_1 c_2' \lambda^8 c_2 \right) \tag{5.3}$$

$$f = \left(c_1' \lambda^3 c_1 c_2' \lambda^3 c_2 + c_1' \lambda^8 c_1 c_2' \lambda^8 c_2 \right) = \frac{-1}{3} \tag{5.4}$$

f is called the color factor. This leads to interaction potential:

$$V = \frac{-1}{3} \frac{\alpha_s \hbar c}{r} \tag{5.5}$$

Observe the potential is repulsive. The quarks in a meson are in fact in the singlet state $\frac{r\bar{r}+g\bar{g}+b\bar{b}}{\sqrt{3}}$. Let us calculate the color factor for this the singlet state. Now, we get a color factor we have to calculate $r\bar{r}$ scattering to $r\bar{r}$ and $r\bar{r}$ scattering to $b\bar{b}$ and $g\bar{g}$. The resulting factor is $\frac{-8}{3}$.

This leads to interaction potential:

$$V = -\frac{8}{3} \frac{\alpha_s \hbar c}{r} \tag{5.6}$$

Observe the potential is attractive explaining why mesons are in singlet state.

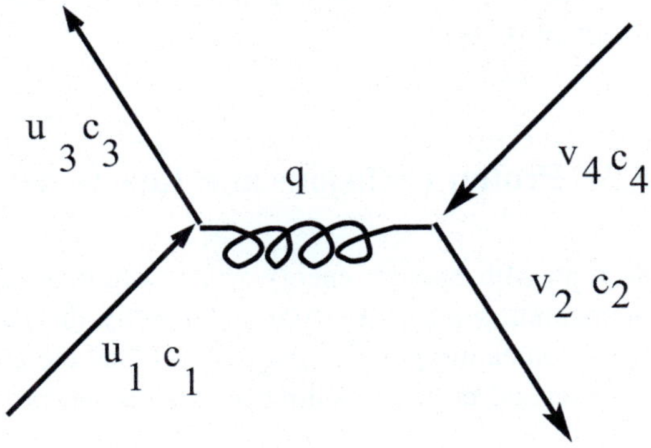

Fig. 5.1 Figure shows quark-antiquark interaction by gluon exchange

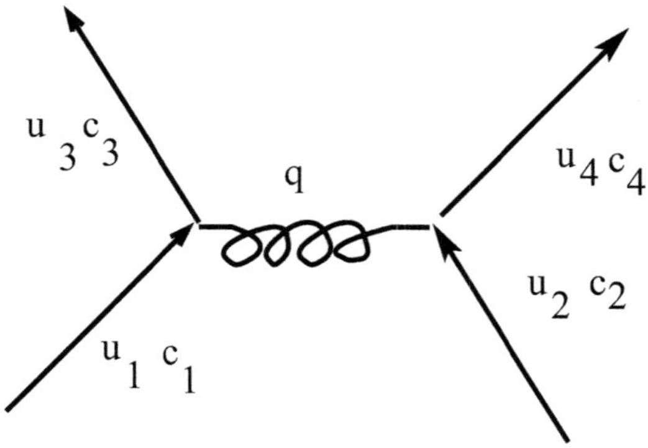

Fig. 5.2 Figure shows quark-quark interaction with gluon exchange

5.2.2 Triplet State

Let the quark-antiquark pair be in the triplet state $\frac{r\bar{r}+b\bar{b}-2g\bar{g}}{\sqrt{6}}$; let us calculate the color factor for this singlet state. Now, we get a color factor we have to calculate $r\bar{r}$ scattering to $r\bar{r}$ and $r\bar{r}$ scattering to $b\bar{b}$ and $g\bar{g}$. The resulting factor is $\frac{-}{3}$.

This leads to interaction potential:

$$V = -\frac{1}{3}\frac{\alpha_s \hbar c}{r} \tag{5.7}$$

5.2.3 Quark-Quark Interaction

For baryons like proton, we have the color singlet state:

$$\frac{rgb - grb + gbr - bgr + brg - rbg}{\sqrt{6}}$$

. We now have quark-quark interaction as shown in Fig. 5.2. Again, interaction potential is negative.

5.3 Proton Collisions and Reactions: Pions and Kaons

Protons with enough energy when colliding can produce pions. They exchange momentum $p+k$ with gluons, not directly, the gluons create an antiquark-quark pair with momentum p and k, respectively, and the quark further exchanges momentum with second proton, resulting in a pion and protons with reduced energy. This is

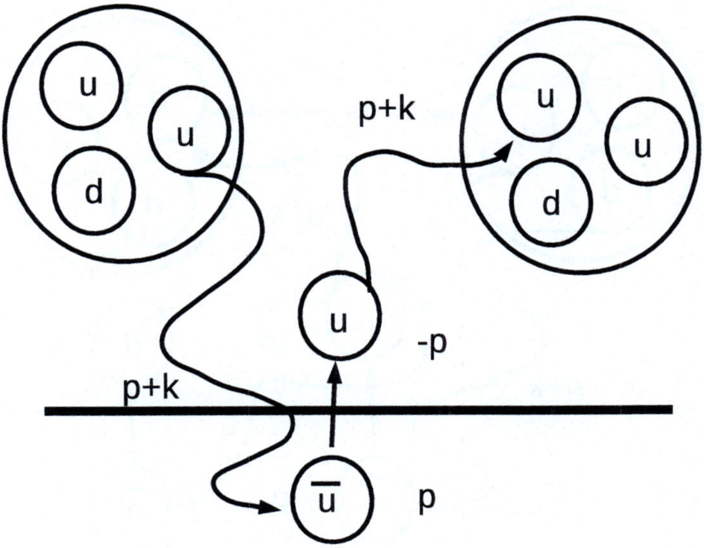

Fig. 5.3 Figure shows how two protons collide with sufficient energy and exchange a $p + k$ momentum photon, not directly, but by exciting a quark-antiquark pair, the neutral pion, the energy of quark pair is paid by change in momentum of protons

shown in Fig. 5.3. The color of the proton quark changes which can be restored by exchanging a third gluon shown in dotted Fig. 5.3.

Protons with enough energy when colliding can produce positive-negative pion pairs. They exchange momentum $p + k$ with gluons, not directly, the gluons create an antiquark-quark pair with momentum p and k, respectively, the quark further exchanges momentum with second proton, and we can repeat the process resulting in two pairs $u\bar{u}$ and $d\bar{d}$ which reassemble to give positive pion π^+ $(u\bar{d})$ and negative pion π^+ $(u\bar{d})$. This is shown in Fig. 5.4.

Protons with enough energy when colliding can produce positive-negative pion pairs. They exchange momentum $p + k$ with gluons, not directly, the gluons create an antiquark-quark pair with momentum p and k, respectively, the quark further exchanges momentum with second proton, and we can repeat the process resulting in two pairs $u\bar{u}$ and $d\bar{d}$ which reassemble to give positive pion K^+ $(u\bar{s})$ and negative kaon K^- $(\bar{u}s)$. This is shown in Fig. 5.5.

5.4 Strong Nuclear Force

Neighboring protons have quarks which can talk through gluons, but proton is color neutral so there is no net potential due to this (Fig. 5.6). But by exchanging gluons, we can create a pion which can exchange momentum between protons. Shown in

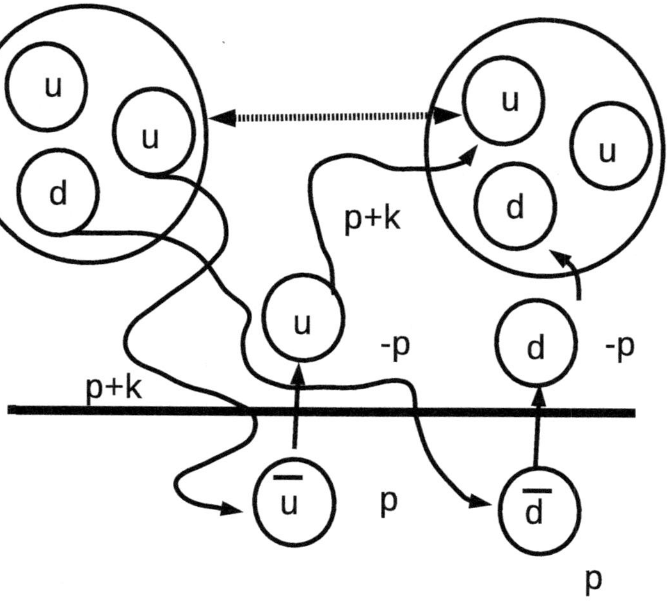

Fig. 5.4 Figure shows how two protons collide with sufficient energy and exchange a $p + k$ momentum photon, not directly, but by exciting two quark-antiquark pairs, the positive and negative pion, the energy of quark pair is paid by change in momentum of protons

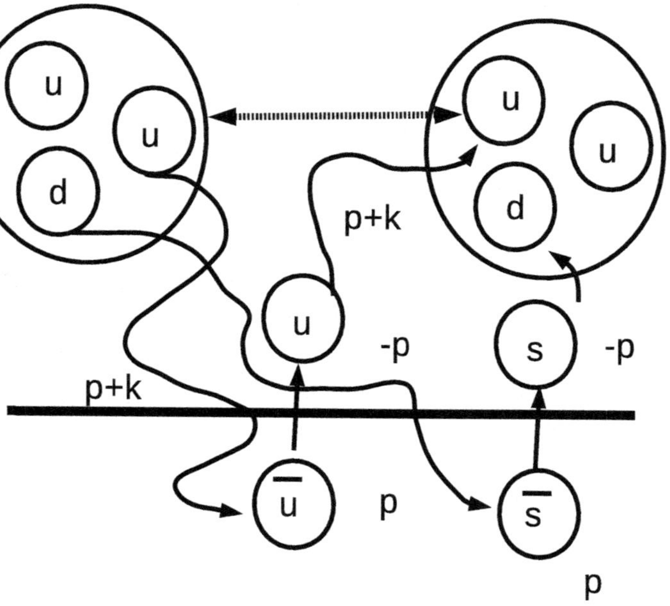

Fig. 5.5 Figure shows how two protons collide with sufficient energy and exchange a $p + k$ momentum photon, not directly, but by exciting two quark-antiquark pairs, the positive and negative kaon, the energy of quark pair is paid by change in momentum of protons

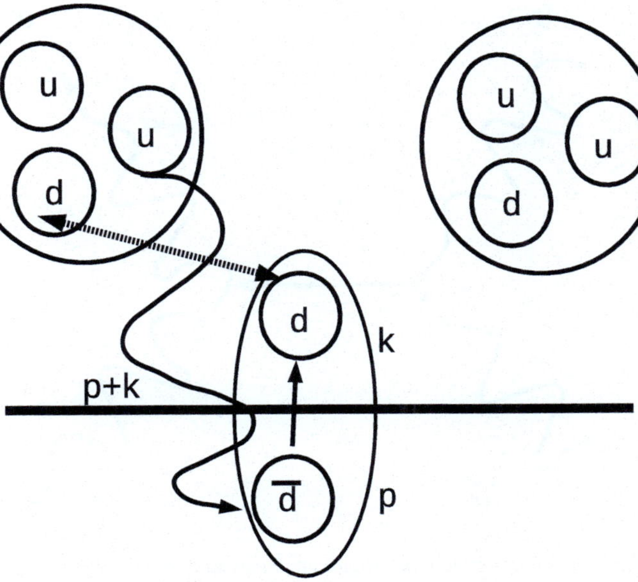

Fig. 5.6 Figure shows how two protons exchange momentum $p + k$, by emitting and absorbing a pion that mediates a nuclear force

Fig. 5.7 is how a u blue quark emits a $p + k$ gluon which excites a antiquark-quark pair with momentum p and k and color $\bar{g}$ and b. Color of the emitting proton quark changes to green, but the produced quark b can exchange again with other proton quarks to make it blue as shown in Fig. 5.7 to produce a quark-antiquark pair that is $g\bar{g}$.

The amplitude of pion exchange $\mathcal{M} \propto \frac{1}{q^2}$, where q is the exchange momentum, which is

$$\mathcal{M} \propto \frac{1}{|k|^2 + m_0^2}$$

, this gives a potential

$$V \propto \frac{exp(-\frac{r}{r_0})}{r},$$

where $r_0 = \frac{\hbar}{m_0 c}$ is the pion Compton wavelength $\sim$ fm, then the potential is very short range around fm scale. At around 2 fm, it is 30 MeV. It is attractive.

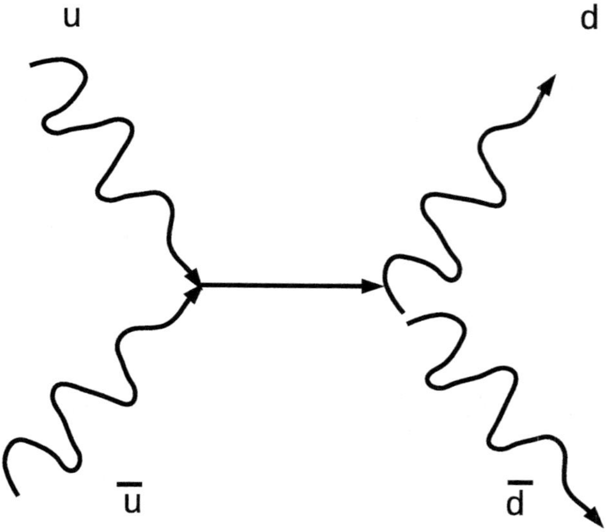

Fig. 5.7 Figure shows how quark-antiquark $u\bar{u}$ annihilate to create the pair $d\bar{d}$

5.5 Pair Production

Pion is $\pi_0 = \frac{u\bar{u}-d\bar{d}}{\sqrt{2}}$; what does it mean? The quark pair $u\bar{u}$ can annihilate to give a gluon which creates pair $d\bar{d}$, called pair production. The two quark-antiquark states are almost degenerate so we can have a superposition $\pi_0 = \frac{u\bar{u}-d\bar{d}}{\sqrt{2}}$, with smaller energy.

5.6 Asymptotic Freedom

In vacuum polarization in QED, we saw how energy of an exchange photon is modified. Same happens to exchange gluon, in QCD. There are now two kinds of vertices as shown in Fig. 5.8, where exchange gluon splits into two gluons (boson vertex) or quark-antiquark pair (fermion vertex). We already study fermion vertex in QED in Chap. 3 and boson vertex in gauge potential in weak interaction in Chap. 4. We saw amplitude of vertex B in Fig. 5.8 is negative, QED correction to photon energy is negative, and interactions become stronger. We saw amplitude of vertex A in Fig. 5.8 is positive, QCD correction to photon energy is positive, and interactions become weaker. The boson vertex interaction scales $\frac{E_1^2}{E_2}E_3$, if we look at small distances, E_1 is large, and hence, we have very weak interactions termed asymptotic freedom.

1. What is quark-antiquark potential at distance of 1 fm if they are in singlet state?

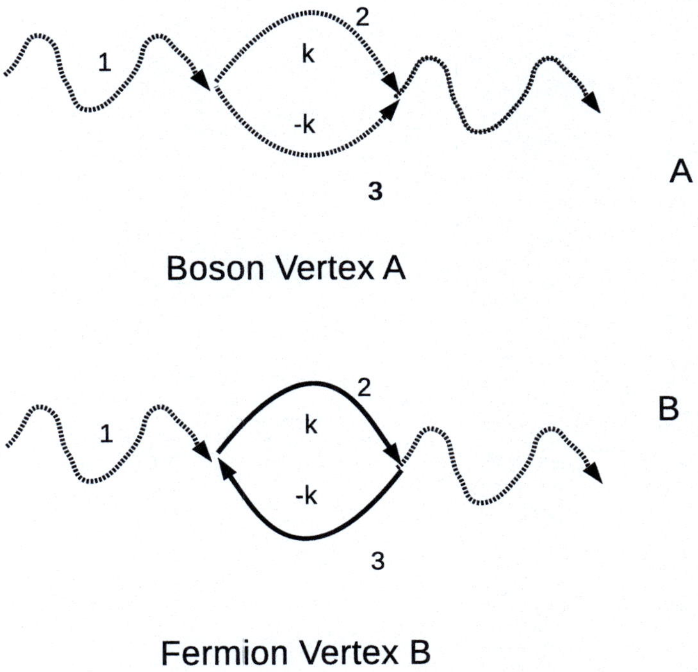

Fig. 5.8 Figure shows the boson and the fermion vertex for energy correction of a gluon

2. What is quark-antiquark potential at distance of 1 fm if they are in triplet state?
3. What is energy of outgoing protons if they collide at 2 GeV each to make a π_0 at rest?
4. What is their outgoing momentum in above?
5. What is energy of outgoing protons if they collide at 2 GeV each to make a π^+ and π^- at rest?

Chapter 6
Bound States and Quantum Chromodynamics

6.1 Quantum Mechanics

6.1.1 Schrödinger Equation

In classical mechanics, we talk about a particle say an electron with a position x and velocity v. In quantum mechanics, particle state is represented by complex waves $\exp(ikx)$ or sum of such waves $\sum_j \exp(ik_j x)$. In complex wave $\exp(ikx)$, k is the wavenumber of the particle. The wave evolves in time as $\exp(i(kx - \omega(k)t))$, $\omega(k)$ is the frequency of the wave and depends on wavenumber k. The dependence $\omega(k)$ is called the dispersion relation of the wave. First postulate of quantum mechanics is that the energy of the wave is $E = \hbar\omega(k)$, where $\hbar$ is a fundamental constant called Planck's constant. Its units are angular momentum and in SI units its value is 6.6×10^{-34}.

The momentum of our complex wave ω, k is simply $\hbar k$.

Now, from classical mechanics, $E = \frac{p^2}{2m}$. Then, we get $\hbar\omega = \frac{\hbar^2 k^2}{2m}$ or $\omega = \frac{\hbar k^2}{2m}$. Thus, my complex wave $\psi(x, t) = \exp(i(kx - \omega t))$ satisfies

$$i\hbar \frac{\partial \psi}{\partial t} = (-\frac{\hbar^2}{2m}\frac{\partial^2 \psi}{\partial x^2}). \tag{6.1}$$

Equation 6.1 is called *Schröedinger* equation. It is still true if we have

$$\psi(x, t) = \sum_j \alpha_j \exp(i(k_j x - \omega(k_j)t)).$$

as individual exponential satisfies these equation.

$\psi(x, t)$ is called a wavefunction of electron; it is superposition of plane waves. This is a feature of quantum mechanics; we can be in superposition of states. It satisfies the Schrödinger equation. All we are saying is that if we start with initial

N. Khaneja, *High Energy Physics*, https://doi.org/10.1007/978-3-031-75645-0_6

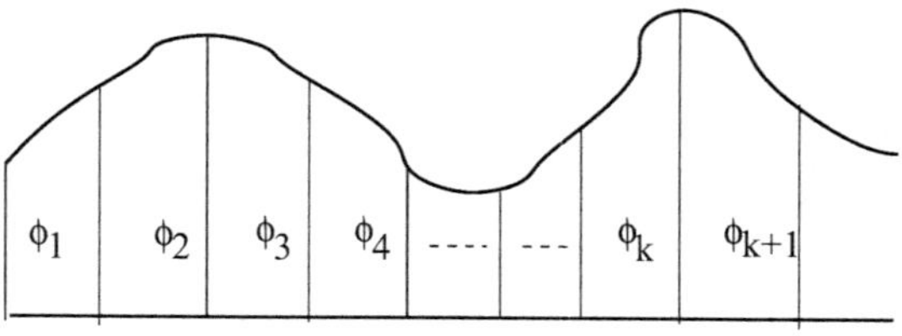

Fig. 6.1 Figure shows how $V(x)$ is decomposed as piecewise constant potential

state $\psi(x) = \sum_j \alpha_j \exp(ik_jx)$, these ways will evolve by their characteristic energies as $\psi(x,t) = \sum_j \alpha_j \exp(i(k_jx - \omega(k_j)t))$ and $\psi(x,t)$ satisfies the Schrödinger equation.

Now, how does my wavefunction evolve if I have a potential V? Then, from classical mechanics $E - V = \frac{p^2}{2m}$, implying $\hbar\omega - V = \frac{\hbar^2 k^2}{2m}$ or my wave satisfies

$$i\hbar\frac{\partial\psi}{\partial t} = (-\hbar\frac{\partial^2}{\partial x^2} + V)\psi. \tag{6.2}$$

and again same is true if we have superposition of plane waves.

Now, how does the evolution of $\psi(x)$ take place when we have $V(x)$? Then, we can break $\psi(x)$ into small pieces ϕ_i over which $V(x)$ is constant as V_i. See Fig. 6.1. Then, each ϕ_i sees a potential V_i. Its evolution will be same if V_i was globally true. Then, we can break ϕ into exponentials and conclude it satisfies the equation:

$$i\hbar\frac{\partial\phi_i}{\partial t} = (-\hbar\frac{\partial^2}{\partial x^2} + V_i)\phi_i. \tag{6.3}$$

Then, adding them all, we get

$$i\hbar\frac{\partial\psi}{\partial t} = (-\hbar\frac{\partial^2}{\partial x^2} + V(x))\psi. \tag{6.4}$$

Thus, we have derived a fundamental equation of quantum mechanics. Wavefunction $\psi(x)$ has a probabilistic interpretation. $\int_a^b |\psi(x)|^2 dx$ gives the probability of finding the particles in the interval $[a, b]$.

6.1.2 Hydrogen Atom

In polar coordinates, $r = \sqrt{x^2 + y^2}$ and $\phi = \tan^{-1}(\frac{y}{x})$.

Then

$$\frac{\partial}{\partial x} = \frac{\partial r}{\partial x}\frac{\partial}{\partial r} + \frac{\partial\phi}{\partial x}\frac{\partial}{\partial\phi} = \cos\phi\frac{\partial}{\partial r} - \frac{\sin\phi}{r}\frac{\partial}{\partial\phi}$$

$$\frac{\partial}{\partial y} = \frac{\partial r}{\partial y}\frac{\partial}{\partial r} + \frac{\partial \phi}{\partial y}\frac{\partial}{\partial \phi} = \sin\phi\frac{\partial}{\partial r} + \frac{\cos\phi}{r}\frac{\partial}{\partial \phi}.$$

$$\frac{\partial^2}{\partial x^2} + \frac{\partial^2}{\partial y^2} = \frac{\partial^2}{\partial r^2} + \frac{1}{r}\frac{\partial}{\partial r} + \frac{1}{r^2}\frac{\partial}{\partial \phi^2}.$$

$$\frac{\partial^2}{\partial z^2} + \frac{\partial^2}{\partial x^2} + \frac{\partial^2}{\partial y^2} = \frac{\partial^2}{\partial z^2} + \frac{\partial^2}{\partial r^2} + \frac{1}{r}\frac{\partial}{\partial r} + \frac{1}{r^2}\frac{\partial}{\partial \phi^2}.$$

Using $R = \sqrt{z^2 + r^2}$ and $\theta = \tan^{-1}(\frac{r}{z})$.

$$\begin{aligned}\frac{\partial^2}{\partial z^2} + \frac{\partial^2}{\partial x^2} + \frac{\partial^2}{\partial y^2} &= \frac{\partial^2}{\partial R^2} + \frac{2}{R}\frac{\partial}{\partial R} + \frac{1}{R^2}\frac{\partial}{\partial \theta^2} + \frac{\cot\theta}{R^2}\frac{\partial}{\partial \theta} + \frac{1}{R^2\sin^2\theta}\frac{\partial}{\partial \phi^2} \\ &= \frac{1}{R^2}\frac{\partial}{\partial R}(R^2\frac{\partial}{\partial R}) + \frac{1}{R^2\sin\theta}\frac{\partial}{\partial \theta}(\sin\theta\frac{\partial}{\partial \theta}) + \frac{1}{R^2\sin^2\theta}\frac{\partial}{\partial \phi^2}.\end{aligned}$$

To Schrodinger eigenvalue, Eq. reads

$$\{\frac{\hbar^2}{2m}(\frac{1}{R^2}\frac{\partial}{\partial R}(R^2\frac{\partial}{\partial R}) + \frac{1}{R^2\sin\theta}\frac{\partial}{\partial \theta}(\sin\theta\frac{\partial}{\partial \theta}) + \frac{1}{R^2\sin^2\theta}\frac{\partial}{\partial \phi^2}) + (E - V(R))\}\psi = 0.$$

$$\{(\frac{\partial}{\partial R}(R^2\frac{\partial}{\partial R}) + \frac{1}{\sin\theta}\frac{\partial}{\partial \theta}(\sin\theta\frac{\partial}{\partial \theta}) + \frac{1}{\sin^2\theta}\frac{\partial}{\partial \phi^2}) + \frac{2mR^2}{\hbar^2}(E - V(R))\}\psi = 0.$$

We write the solution $\psi = f(R)Y(\theta, \phi)$:

$$(\frac{1}{\sin\theta}\frac{\partial}{\partial \theta}(\sin\theta\frac{\partial}{\partial \theta}) + \frac{1}{\sin^2\theta}\frac{\partial}{\partial \phi^2} + \underbrace{l(l+1)}_{E_1})Y(\theta, \phi) = 0. \tag{6.5}$$

Writing $Y(\theta, \phi) = \Theta(\theta)e^{im\phi}$, we get

$$(\frac{1}{\sin\theta}\frac{\partial}{\partial \theta}(\sin\theta\frac{\partial}{\partial \theta}) - \frac{1}{\sin^2\theta}m^2 + \underbrace{l(l+1)}_{E_1})\Theta(\theta) = 0.$$

For $x = \cos\theta$, the above equation reads

$$(1 - x^2)\Theta'' - 2x\Theta' + (l(l+1) - \frac{m^2}{1 - x^2})\Theta = 0.$$

The solution Θ_l^m exits for integer l, m satisfying $0 \leq |m| \leq l$. For $m \geq 0$

$$\Theta_l^m(x) = \frac{(-1)^m}{2^l l!}(1-x^2)^{\frac{m}{2}} \frac{d^{l+m}}{dx^{l+m}}(x^2-1)^l.$$

with

$$\Theta_l^{-m}(x) = (-1)^m \frac{(l-m)!}{(l+m)!}\Theta_l^m(x).$$

Then, the equation for R gives

$$\frac{\partial}{\partial R}(R^2 \frac{\partial f}{\partial R}) = (l(l+1) + \frac{2mR^2}{\hbar^2}(V(R)-E))f.$$

Let $u = Rf$, then

$$-\frac{\hbar^2}{2m}\frac{\partial^2 u}{\partial R^2} + (V + \frac{\hbar^2}{2mR^2}l(l+1))u = Eu,$$

where $V = \frac{-e^2}{4\pi\epsilon_0 r}$. This is one-dimensional Schrodinger equation. Guess a solution of the form $u(r) = R^{l+1}e^{-\frac{R}{a_0}}$. Then, twice differentiating R^{l+1} cancels the centrifugal part. Differentiating R^{l+1} and $e^{-\frac{R}{a_0}}$ cancels V, when $\frac{\hbar^2}{m}\frac{l+1}{a_0} = \frac{e^2}{4\pi\epsilon_0}$, i.e.,

$$a_0 = \frac{(l+1)\hbar^2 4\pi\epsilon_0}{me^2}, \quad E = \frac{\hbar^2}{2ma_0^2}.$$

However, we do not have to cancel V immediately. We can add another term:

$$u(r) = R^{l+1}e^{-\frac{R}{a_0}} + c_1 R^{l+2}e^{-\frac{R}{a_0}}.$$

Then, centrifugal part of second term c_1 can cancel the part of first term obtained by differentiating R^{l+1} and $e^{-\frac{R}{a_0}}$. For this, c_1 has to be chosen correct. Now, we cancel V by differentiating R^{l+2} and $e^{-\frac{R}{a_0}}$.

Then, in general

$$u(r) = R^{l+1}e^{-\frac{R}{a_0}}(1 + \sum_{j=1}^{d} c_j R^j),$$

with $n = l + d + 1$, the principle quantum number. Then

$$\frac{\hbar^2}{m}\frac{n}{a_0} = \frac{e^2}{4\pi\epsilon_0}. \quad a_0 \propto n$$

and

$$\frac{c_j}{c_{j-1}} = \frac{2(l + j - n)a_0^{-1}}{j(2l + j + 1)}.$$

This gives a_0 and finally

$$E = \frac{\hbar^2}{2ma_0^2}, \quad E \propto \frac{1}{n^2}.$$

6.1.3 Angular Momentum

$$L = r \times p.$$

$$L_x = yp_Z - zp_y, \quad L_y = zp_x - xp_z, \quad L_z = xp_y - yp_x.$$

Using $[p_x, x] = -i\hbar$, etc., we have

$$L^2 = L_x^2 + L_y^2 + L_z^2 = R^2(p_x^2 + p_y^2 + p_z^2) - (xp_x + yp_y + zp_z - i\hbar)^2 + \hbar^2.$$

A quick calculation shows

$$xp_x + yp_y + zp_z = -i\hbar R\frac{\partial}{\partial R}.$$

Now, substituting for

$$p_x^2 + p_y^2 + p_z^2 = -\hbar^2\left(\frac{1}{R^2}\frac{\partial}{\partial R}\left(R^2\frac{\partial}{\partial R}\right) + \frac{1}{R^2\sin\theta}\frac{\partial}{\partial\theta}\left(\sin\theta\frac{\partial}{\partial\theta}\right) + \frac{1}{R^2\sin^2\theta}\frac{\partial}{\partial\phi^2}\right) \tag{6.6}$$

$$L^2 = -\hbar^2\left(\frac{1}{\sin\theta}\frac{\partial}{\partial\theta}\left(\sin\theta\frac{\partial}{\partial\theta}\right) + \frac{1}{R^2\sin^2\theta}\frac{\partial}{\partial\phi^2}\right). \tag{6.7}$$

Then, from Eq. 6.5

$$L^2Y(\theta, \phi) = \hbar^2 l(l + 1).$$

and

$$L_z Y(\theta, \phi) = -i\hbar \frac{\partial}{\partial \phi} Y(\theta, \phi) = \hbar m Y(\theta, \phi).$$

We denote the eigenfunction as Y_{lm}.
Observe easily verifiable commutation relations:

$$[L_x, L_y] = i\hbar L_z, \quad [L_y, L_z] = i\hbar L_x, \quad [L_z, L_x] = i\hbar L_y. \tag{6.8}$$

Define $L^- = L_x - iL_y$ and $L^+ = L_x + iL_y$. L^- is called lowering operator and L^+ is called raising operator:

$$[L^2, L^\pm] = 0, \quad [L_z, L^\pm] = \pm\hbar L^\pm. \tag{6.9}$$

Then, note $[L^2, L^\pm]Y_{lm} = 0$ implies $L^2 L^\pm Y_{lm} = \hbar^2 l(l+1) L^\pm Y_{lm}$; hence, $L^\pm Y_{lm}$ is a linear combination of Y_{lm} for different m. Now, $[L_z, L^+]Y_{lm} = L^+ Y_{lm}$ implying $L_z L^+ Y_{lm} = \hbar(m+1) L^+ Y_{lm}$ implying $L^+ Y_{lm} = a_m Y_{l,m+1}$. Similarly, $L^- Y_{lm} = b_m Y_{l,m-1}$. Then, observe $L^+ Y_{ll} = 0$ and $L^- Y_{l,-l} = 0$. Furthermore

$$[L^+, L^-] = 2\hbar L_z. \tag{6.10}$$

Furthermore, we get

$$L^+ L^- + L^- L^+ = 2(L^2 - L_z^2). \tag{6.11}$$

Then, we get

$$L^+ L^- = L^2 - L_z^2 + \hbar L_z \tag{6.12}$$

$$L^- L^+ = L^2 - L_z^2 - \hbar L_z. \tag{6.13}$$

For normalized Y_{lm}, we get

$$b_{lm} = \hbar\sqrt{l(l+1) - m(m-1)} \tag{6.14}$$

$$a_{lm} = \hbar\sqrt{l(l+1) - m(m+1)} \tag{6.15}$$

We talked about orbitals with principle quantum number n and integer angular momentum number l and z angular momentum l, with $|m| \leq l \leq n-1$. Here, l was integer. In principle, it can be half integer and is ascribed to an intrinsic angular momentum called spin (Fig. 6.2). We use the quantum number s instead of l. In particular, $s = \frac{1}{2}$ is called spin $\frac{1}{2}$ a property of electron. We then have two values of $s_z = \pm\frac{1}{2}$. Then, an electron has two sets of quantum numbers l, m and s, s_z.

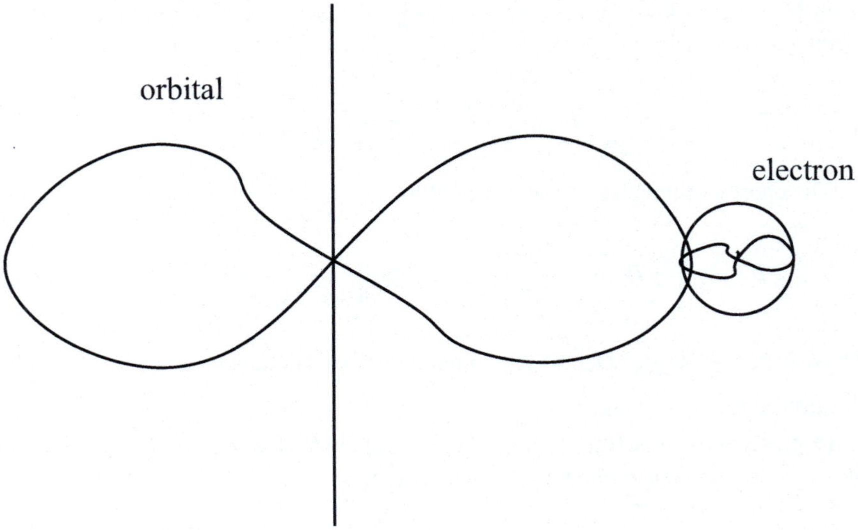

Fig. 6.2 Figure shows an atomic orbital and an electron with an inner orbital that constitutes its spin angular momentum

6.2 Fine Structure and Spin Orbital Coupling

We talked about spin. Let's try to understand the physics of it. You are familiar with earth spinning on its axis. This gives earth a angular momentum. Now, imagine our earth was charged. Then, spinning will give earth a magnetic moment. Imagine a loop of wire carrying current (circulating charge), and then, it has a magnetic moment $M = I.A$, where I is the current and A area of the loop, from your basic physics. Now, imagine a charge q going around in a loop of radius r, with angular velocity ω. Then, it makes $\frac{\omega}{2\pi}$ rotations per sec. The current is then $\frac{q\omega}{2\pi}$ and its magnetic moment is $\mu_S = \frac{q\omega\pi r^2}{2\pi} = \frac{q}{2m}(mvr)$ where $l = mvr$ is the angular momentum. Then, $\mu_S = \frac{q}{2m}L$, the ratio $\gamma = \frac{q}{2m}$ is called the gyromagnetic ratio, and it relates angular momentum to magnetic moment. For reasons coming from relativity, we in fact have $\gamma = \frac{q}{m}$.

There is coupling between electron spin and orbital angular momentum. There is coupling Hamiltonian of the form:

$$H_{so} = \beta L \cdot S. \tag{6.16}$$

Let us see how this coupling arises. When electron is at a certain point on its orbital, it has a velocity v and momentum p. From perspective of the electron, the nucleus is moving in the opposite direction with same magnitude of velocity. Then, from Biot-Savart law, the moving nucleus produces a magnetic field on the site of electron given by

$$B = \frac{e\mu_0}{4\pi}\frac{p \times r}{mr^3} = \frac{e\mu_0}{4\pi}\frac{L}{mr^3}. \tag{6.17}$$

The energy of the electron in this field is

$$B \cdot \mu_S = \gamma B \cdot S = \frac{e^2\mu_0}{4\pi m^2 r^3} L \cdot S = \frac{e^2}{4\pi\epsilon_0 c^2 m^2 r^3} L \cdot S. \tag{6.18}$$

Thus, $\beta = \frac{e^2}{4\pi\epsilon_0 c^2 m^2 r^3}$. Due to phenomenon called Thomas precession, β is called by another factor of 2 and $\beta = \frac{e^2}{8\pi\epsilon_0 c^2 m^2 r^3}$.

In presence of this Hamiltonian, our orbitals will change. Let us compute how the orbitals change and what are the new energies:

$$L \cdot S = L_z S_z + L_x S_x + L_y S_y = L_z S_z + \frac{L^+ S^- + L^- S^+}{2}. \tag{6.19}$$

For this, define a new operator:

$$J^2 = (L + S)^2 = L^2 + S^2 + 2L \cdot S. \tag{6.20}$$

$$J_z = L_z + S_z, \quad J^{\pm} = L^{\pm} + S^{\pm}. \tag{6.21}$$

Given l and s, we start with the state $l_z = l$ and $s_z = s$. Denote this state by (l, s). This state is an eigenstate of the operator $L \cdot S$ with eigenvalue l, s, and hence, it is an eigenstate of J^2 with eigenvalue $j(j+1)$ with $j = l + s$. Now, as before, we can apply lowering operator. From the last section, $J^-(j, j_z) = b(j, j_{z-1})$ with $b = \hbar\sqrt{j(j+1) - j_z(j_z - 1)}$, so by applications of J^- we decrease j_z until it is $-j$. Hence, we have constructed $2j$ or $2j + 1$ orbitals depending on if j is integer or half integer.

Observe $J^-(l, s) = (l - 1, s) + (l, s - 1)$. There is another orthogonal state $e_1 = (l - 1, s) - (l, s - 1)$ which is eigenfunction of J_z with eigenvalue $l + s - 1$ and hence must be an eigenfunction of J^2. The eigenvalue of J^2 cannot be $j(j+1)$ as we have exhausted all these vectors as $J^+ e_1 = 0$. Only possible value of J^2 is $(j-1)j$, we again apply lowering operators and go from $j_z = j - 1, \ldots, -(j-1)$.

Now, we consider $J^{-2}(l, s) = (l - 2, s) + (l - 1, s - 1) + (l, s - 2)$, which has $J_z = l + s - 2$. We have constructed two eigenvectors $J^2 = j(j+1)$ and $J^2 = (j-1)j$. We can form a third eigenvector, we can show it has J^2 value $(j-1)(j-2)$, and we can again apply lowering operators and construct eigenvectors with J^2. Instead of writing J^2, we say this J which in this case has value $j - 2$.

We start with one term (l, s). Then, $J^-(l, s)$ has two terms, and$J^{-2}(l, s)$ has three terms. This process continues till smaller of l, s say s becomes $-s$. Then,

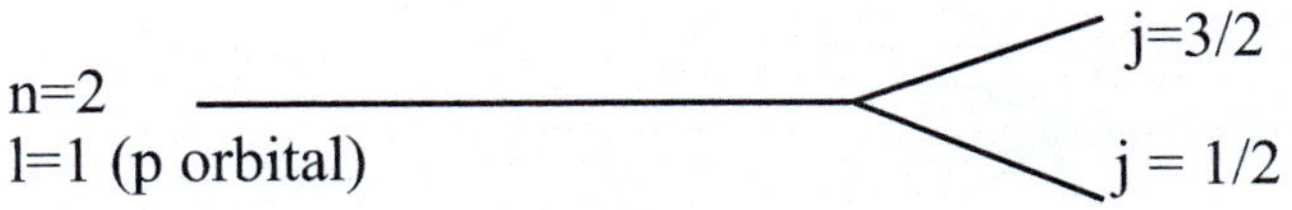

Fig. 6.3 Figure shows how $n = 2$, p orbital gets split due to fine structure

lowering doesn't increase number of terms. Then, starting with $j = l + s$, we go until $j = l - s$. Thus, all states can be indexed by $j = l + s, \ldots, l - s$, and for a given j, we have $j_z = j, \ldots, -j$. Thus, starting with state $|l, l_z\rangle|s, s - z\rangle$, we have formed state:

$$|j, j_z\rangle = \sum_{l_z, s_z} c_{l_z, s_z} |l, l_z\rangle |s, s_z\rangle, \tag{6.22}$$

where as just told, $j = l + s, \ldots, l - s$ and for a given j we have $j_z = j, \ldots, -j$.

In the basis $|j, j_z\rangle$, we have $L \cdot S$ is diagonal with eigenvalue $\frac{j(j+1) - l(l+1) - s(s+1)}{2}$. The coefficients c_{l_z, s_z} are called Clebsch-Gordan coefficients. Figure 6.3 shows how $n = 2$, p orbital gets split due to fine structure.

As we can see in the figure, an energy level $n = 1, l = 1$ in presence of $L \cdot S$ coupling gets split into two set of orbitals $j = \frac{3}{2}$ with $(j_z = \frac{3}{2}, \ldots, -\frac{3}{2})$ and $j = \frac{1}{2}$ with $(j_z = \frac{1}{2}, \ldots, -\frac{1}{2})$ with different energies. This is called *fine structure*. If we estimate how big this is, it is $\beta = \frac{\hbar^2 e^2}{4\pi c^2 \epsilon_0 m_e^2 r^3} \sim 10^2$ eV $\sim 10^3$ GHz. It arises because the angular momentum of the orbital and the spin of the electron talk to each other. Evaluating spin orbit coupling

$$\langle \frac{1}{r^3} \rangle = \frac{1}{n^3 l(l + \frac{1}{2})(l + 1) a^3} \tag{6.23}$$

$$E_{so} = \alpha^4 mc^2 \frac{j(j + 1) - l(l + 1) - \frac{3}{4}}{4n^3 l(l + \frac{1}{2})(l + 1)} \tag{6.24}$$

Electron has a spin, so does the nucleus of the atom. It is called nuclear spin. We denote nuclear spin with I like we denote electron spin with S. We assume that we again have an interaction between nuclear spin and electron orbital and spin angular momentum as

$$I \cdot (L + S) = I \cdot J. \tag{6.25}$$

What was between L and S is between I and J so we can define the total angular momentum:

$$F = I + J. \tag{6.26}$$

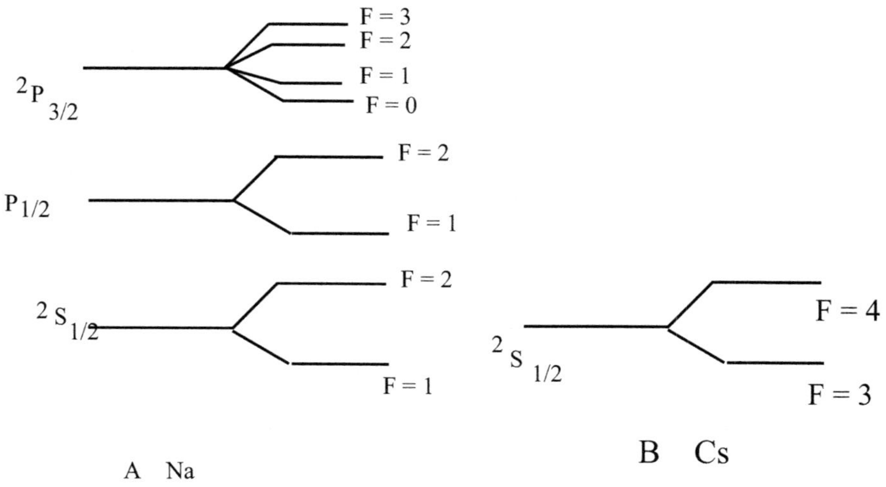

Fig. 6.4 (**a**) shows hyperfine levels for sodium. (**b**) shows hyperfine levels for cesium

Given i and j, the coupling gives f taking values between $i + j, \ldots, |i - j|$. Thus, a j orbital gets split into f orbitals. This is called hyperfine splitting. The eigenvalues of $I \cdot J$ take one values $\frac{f(f+1)-j(j+1)-i(i+1)}{2}$. Thus, if we estimate how much this is, it is $\beta = \frac{\hbar^2 e^2 \mu_0}{4\pi m_e m_p r^3} \sim 1GHz$, where m_p is proton mass which is 10^3 heavier than electron mass (Fig. 6.4).

6.3 Relativistic Correction

In Schrödinger equation, we used kinetic energy as $\frac{p^2}{2m}$. If we use relativistic formula of

$$E = \sqrt{(pc)^2 + (mc^2)^2} \sim mc^2 + \frac{p^2}{2m} - \frac{p^4}{8m^3c^2}. \tag{6.27}$$

Then, we find that

We have correction to energy

$$\Delta E_{rel} = -\frac{p^4}{8m^3c^2}. \tag{6.28}$$

Using the fact $\frac{p^2}{2m} = E - V$, on an orbital, we can calculate

$$\langle \Delta E_{rel} \rangle = -\frac{1}{2mc^2} \; E_n^2 - 2E_n \langle V \rangle + \langle V^2 \rangle. \tag{6.29}$$

$$E_n = -\frac{\alpha^2 mc^2}{2n^2}; \;\; \alpha = \frac{e^2}{4\pi \epsilon_0 \hbar c}. \tag{6.30}$$

Using $V = \frac{e^2}{4\pi \epsilon_0 r}$

$$\langle \frac{1}{r} \rangle = \frac{1}{n^2 a}, \;\; \langle \frac{1}{r^2} \rangle = \frac{1}{n^3 (l + \frac{1}{2}) a^2} \tag{6.31}$$

where a is Bohr radius. Putting everything together, we find

$$\langle \Delta E_{rel} \rangle = -\frac{\alpha^4 mc^2}{4n^4} (\frac{2n}{l + \frac{1}{2}} - \frac{3}{2}). \tag{6.32}$$

Adding Eqs. 6.33 and 6.24, we get what is called fine structure:

$$\langle \Delta E_{fs} \rangle = -\frac{\alpha^4 mc^2}{4n^4} (\frac{2n}{j + \frac{1}{2}} - \frac{3}{2}). \tag{6.33}$$

6.4 Lamb Shift

6.5 Positronium

Positronium (Ps) is a system consisting of an electron and its antiparticle, a positron, bound together into an exotic atom, specifically an onium. The system is unstable: the two particles annihilate each other to predominantly produce two or three gamma rays, depending on the relative spin states. The orbit and energy levels of the two particles are similar to that of the hydrogen atom (which is a bound state of a proton and an electron). However, because of the reduced mass, the frequencies of the spectral lines are less than half of the corresponding hydrogen lines (Fig. 6.5).

6.6 Quarkonium

In positronium, the electron and positron are bound by exchange of a photon. In quarkonium, the quark-antiquark pair $q\bar{q}$ is bound by exchange of a gluon. We have

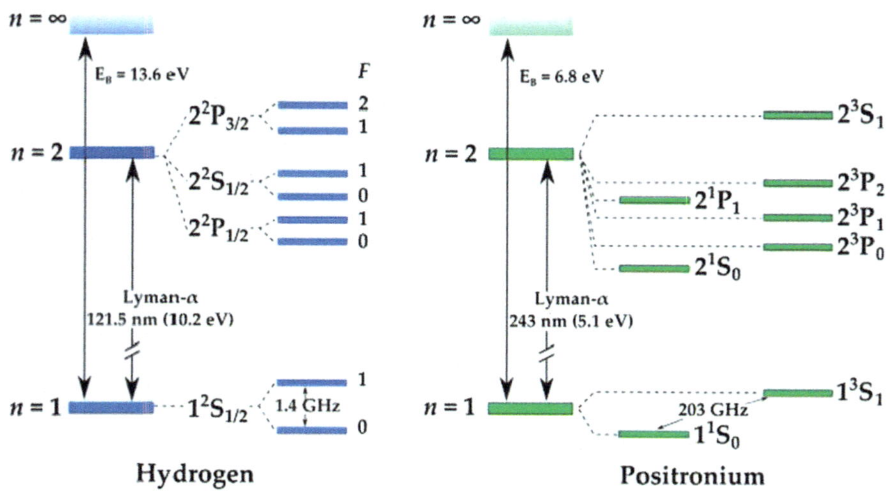

Fig. 6.5 Figure shows energy levels for hydrogen atom and positronium

already seen in Eq. (5.6) that the pair is in singlet state and the binding potential

$$V = -\frac{8}{3}\frac{\alpha_s \hbar c}{r}$$

.

6.7 Proton and Baryons

Proton quark content is uud, but u can scatter to d via weak interaction and emitted W-boson can scatter d to u , so we have the eigenstate $u\frac{ud+du}{\sqrt{2}}$ but again we have a third quark we can scatter to so the completely symmetric state is $\frac{uud+udu+duu}{\sqrt{3}}$. $u\frac{ud-du}{\sqrt{2}}$ is higher energy, but an eigenstate, so possible, it's called anti symmetric combination. If all three quarks are different uds, then of course the completely symmetric state is

$$\frac{uds + dus + dsu + usd + sud + sdu}{\sqrt{6}}.$$

6.8 Proton Spin

Proton has three spin $\frac{1}{2}$, with the spin interaction Hamiltonian

$$H = As_1 \cdot (s_2 + s_3) + Bs_2 \cdot s_3,$$

with $A = -\frac{\mu_0 \hbar^2 \gamma_1 \gamma_2}{a^3}$, where $\gamma_1 = \frac{-e}{3m}$, $\gamma_2 = \gamma_3 = \frac{2e}{3m}$, and a is proton radius. A is positive and B negative. We want $j = s_2 + s_3$ to be spin 1, so that $s_2 \cdot s_3 = j^2 - s_2^2 - s_3^2$ is positive and $f = s_1 + j$ to be spin $\frac{1}{2}$, $s_1 \cdot (s_2 + s_3) = f^2 - j^2 - s_1^2 = -j^2$ is negative. So, that total energy is negative and spin $\frac{1}{2}$.

So, the spin combination is $uud \underbrace{(\uparrow\downarrow\uparrow + \downarrow\uparrow\uparrow - 2\uparrow\uparrow\downarrow)}_{\phi}$ with total spin $\frac{1}{2}$.

What is energy of the spin combination ϕ, in a static field B_0, along z, it is

$$-\phi'(\gamma_2 s_{2z} + \gamma_3 s_{3z} + \gamma_1 s_{1z})\phi \sim B\frac{3e}{m_p}.$$

This means the gyromagnetic ratio is $\sim \frac{3e}{m} \sim 3 \times 10^8$ SI units.

6.9 Meson Spin

The quark-antiquark pair $q\bar{q}$ has each spin $\frac{1}{2}$, with the spin interaction Hamiltonian

$$H = As_1 \cdot s_2,$$

with $A = -\frac{\mu_0 \hbar^2 \gamma_1 \gamma_2}{a^3}$, where $\gamma_1 = \frac{q}{m}$, $\gamma_2 = \frac{\bar{q}}{m}$, and a is proton radius. A is positive as in $u\bar{u}$. We want $j = s_1 + s_2$ to be spin 0, so that $s_1 \cdot s_2 = j^2 - s_1^2 - s_2^2$ is negative.

So, the spin combination is $u\bar{u} \underbrace{(\uparrow\downarrow - \downarrow\uparrow)}_{\phi}$ and has spin $j = 0$.

The excited meson $u\bar{u} \underbrace{(\uparrow\downarrow + \downarrow\uparrow)}_{\phi}$ and has spin $j = 1$ has clearly higher energy.

6.10 CP Violation

The kaons K_0 (d $\bar{s}$) and $\bar{K}_0$ ($\bar{d}$ s) are both not eigenstates of the weak interaction, as weak reactions

$$d\bar{s} \to u\bar{u} \to \bar{d}s$$

say that true eigenstates are $K_S = \frac{K_0 + \bar{K}_0}{\sqrt{2}}$ and $K_L = \frac{K_0 - \bar{K}_0}{\sqrt{2}}$, which decay to two pions or three pions in 10^{-11} s or 10^{-8} s, respectively.

Fitch and Cronin in 1964 prepared a mixture of K_L and K_S that traversed a distance and all K_s decayed early on and only K_L are expected in end, but they found there was still some K_S, in the end, which could only be explained, if K_0 and $\bar{K}_0$ are not same mass. We have

$K_S = \frac{(1-\epsilon)K_0+(1+\epsilon)\bar{K}_0}{\sqrt{2}} = K_s + \epsilon K_L$ and

$K_L = \frac{(1-\epsilon)K_0-(1+\epsilon)\bar{K}_0}{\sqrt{2}} = K_L - \epsilon K_S$ and this small K_s in K_L can explain their experiment. The unequal K_0 and $\bar{K}_0$ masses signify CP violation.

6.11 Problems

1. Spin 1 is coupled to spin $\frac{1}{2}$; what are the possible total angular momentum of coupled spin system?
2. What is the energy of these different momentum states if coupling is $JI \cdot S$?
3. What is the gyromagnetic ratio of a neutron (we calculated proton in the chapter)?
4. What is the gyromagnetic ratio of π_0?
5. What is the gyromagnetic ratio of π^+?

Chapter 7
Collisions: Electron-Proton, Proton-Antiproton, and Proton-Proton

7.1 Electron-Proton Scattering

Beautiful electron-proton scattering experiments were carried out by Robert Hofstadter in the 1950s [1]. These were electron scattering experiments that can be elastic or inelastic (where we excite internal modes of the nucleus). Cross section of scattering sheds light on spatial charge and magnetic moment distribution of the proton (any other nucleus in general).

7.1.1 Elastic Scattering

Let's first discuss elastic scattering [2] where nucleus internal modes are not excited, so that its mass stays same. This is at electron energies $\ll Mc^2$ (in MeV). Figure 7.1 depicts how electron scatters of nucleus at certain choice of θ. Let E_0 and E_1 be incident energy of electron and m mass of electron, M mass of nucleus which is at rest, and then conserving energy momentum gives

$$\frac{1}{E_1} - \frac{1}{E_0} = \frac{1 - \cos\theta}{Mc^2}, \tag{7.1}$$

when electron energies are relativistic with $E_0 = \frac{hc}{\lambda_0}$ and $E_1 = \frac{hc}{\lambda_1}$, with λ_i de Broglie wavelengths, we have

$$\lambda_1 - \lambda_0 = \frac{h(1 - \cos\theta)}{Mc}, \tag{7.2}$$

Proton is three quarks as an approximation equal masses, theta x_1, x_2, x_3 be their coordinates, and let $X_1 = \frac{x_1+x_2+x_3}{3}$, $X_2 = x_1 - x_2$ and $X_3 = \frac{x_1+x_2}{2} - x_3$ be center of mass and two relative coordinates.

N. Khaneja, *High Energy Physics*, https://doi.org/10.1007/978-3-031-75645-0_7

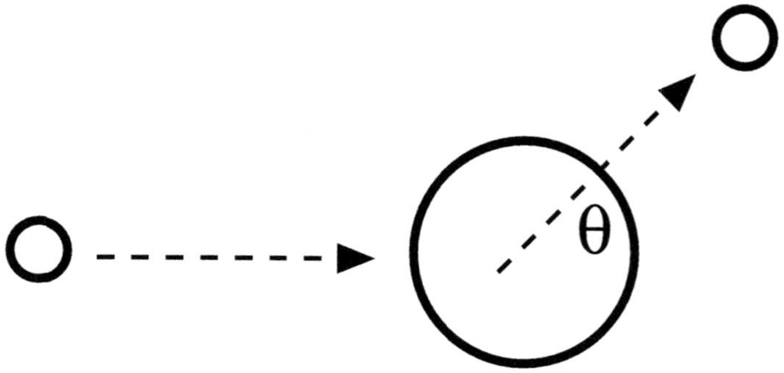

Fig. 7.1 Figure depicts how electron scatters of nucleus at certain choice of θ

Using $\sum_i k_i x_i = \sum_i K_i X_i$, we have $K_1 = k_1 + k_2 + k_3$, $K_2 = \frac{k_1 - k_2}{2}$, and $K_3 = \frac{k_1 + k_2 - 2k_3}{3}$.

When electron transfers momentum q to k_1, we have proton wavefunction ϕ_0 change by

$$\phi_0' = \exp(iqx_1)\phi_0 = \exp(iqX_1)\exp(i\frac{q}{2}X_2)\exp(i\frac{q}{3}X_3)\phi_0$$

The amplitude for this momentum transfer is $\mathcal{M}_0 \propto \frac{ee_1}{q^2\epsilon_0 V}$, where V is electron column and e_1 quark charge. Overlap of new wavefunction with old one is simply

$$\Omega = \langle\phi_0'|\phi_0\rangle = \int \cos(\frac{q}{2}X_2)\phi_0^2 dX_2 \int \cos(\frac{q}{3}X_3)\phi_0^2 dX_3 \sim \frac{1}{q^2}$$

Cross section of scattering $\propto \Omega^2 \propto q^{-4}$, that's it, this is what we find in the experiments, and the elastic cross section dies as $\frac{1}{q^4}$ which means there are **three quarks** ! else it will die as $\frac{1}{q^{2n}}$ for n quarks, and that's it.

Now, let's calculate $\mathcal{M}_0$ in three different regimes when energy E of the incident electron satisfies $E \ll m_e c^2$; this is the limit of elastic scattering.

In this limit, straightforward computation shows for p momentum of the electron, the cross section

$$l \sim \frac{\alpha\hbar}{p\sin^2\frac{\theta}{2}}, \tag{7.3}$$

its scaled de Broglie wavelength. More precisely, we get

$$l \sim \frac{\alpha\hbar\sqrt{1 + (\frac{v}{c}\cos\frac{\theta}{2})^2}}{p\sin^2\frac{\theta}{2}}, \tag{7.4}$$

For incident and scattered electrons at energy E_1 and E_3 and $E_1 \gg m_e c^2$, we have

$$l \sim \frac{\alpha \hbar c f(\theta)}{E_1 \sin^2 \frac{\theta}{2}} \frac{E_3}{E_1}, \tag{7.5}$$

Detail of $f(\theta)$ is a spinor excercise.

7.1.2 Deep Inelastic Scattering

We talked about elastic scattering of electrons and protons. But at high energies, we can have inelastic scattering [3], whereby we can excite the internal modes off the proton, such that its internal energy or mass rises from M to W. Since q needed to create new mass is big comparable r_0^{-1} (radius of proton), cross section will be every small and will die more with increasing q. But we do not need to talk to proton directly. We can first exchange momentum and create a quark-antiquark $f\bar{f}$ pair (meson) as shown in Fig. 7.2, this is creating energy , but we cannot just do it in thin air, because we cannot balance energy; we do this by further exchanging

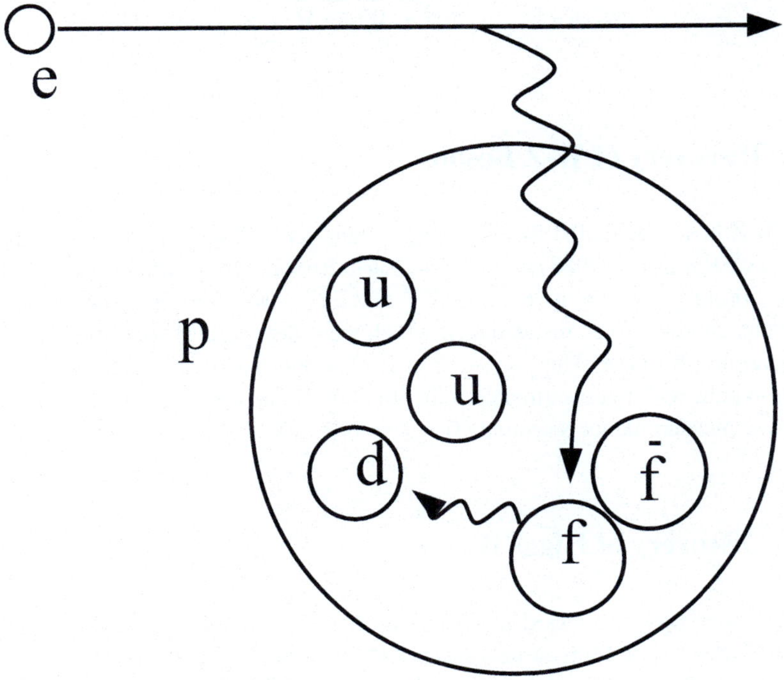

Fig. 7.2 Figure depicts deep inelastic scattering of electron and proton

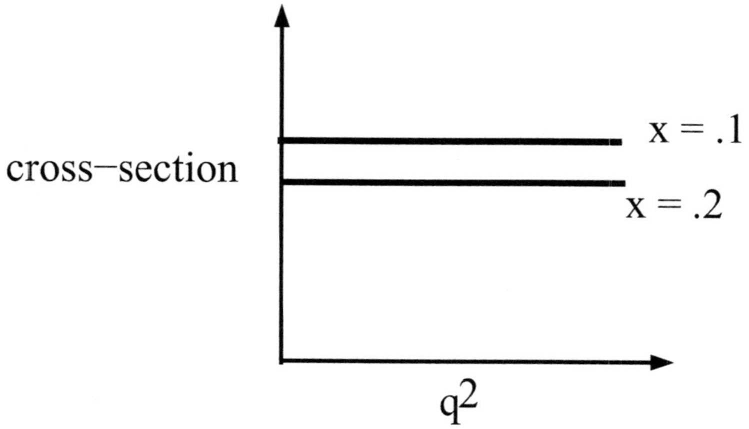

Fig. 7.3 Figure depicts inelastic cross section as a function of q^2 for given x

momentum with the proton and burning this energy in the heavy mass of proton (Fig. 7.3). Scattering amplitude of $f\bar{f}$ has no q dependence, we are just scattering a free particle, so inelastic cross section is independent of q for a gives x, where

$$x = \frac{q^2}{q^2 + W^2 - M^2}$$

7.2 Discovery of W-Z Bosons

The W-Z boson predicted by Weinberg, Salam, and Glasgow (1968) were discovered in CERN lab in (1983) by group of Carlo Rubbia. They collided protons (uud) and antiprotons ($\bar{u}\bar{u}\bar{d}$) to give $u\bar{d} \to W^+$ and $\bar{u}d \to W^-$ bosons, which decayed to positrons and electrons, respectively, which were detected. Masses were estimated to be around 80 GeV. The $u\bar{u} \to Z$ was also detected by its decay to electron-positron pair, with mass estimated to be 90 GeV. Clearly, the center of mass energy for these measurements is around $90 \times 3 \sim 270$ GeV.

7.3 Discovery of Higgs Boson

Higgs boson was discovered at CERN in 2013 in two big projects ATLAS and CIMMS. Protons at Teravolt energies collide to generate top quark-antiquark pairs which recombine to give Higgs boson as in Fig. 7.4. Figure 7.4 shows how each proton produces a Higgs pair, and the two opposite momentum Higgs from colliding

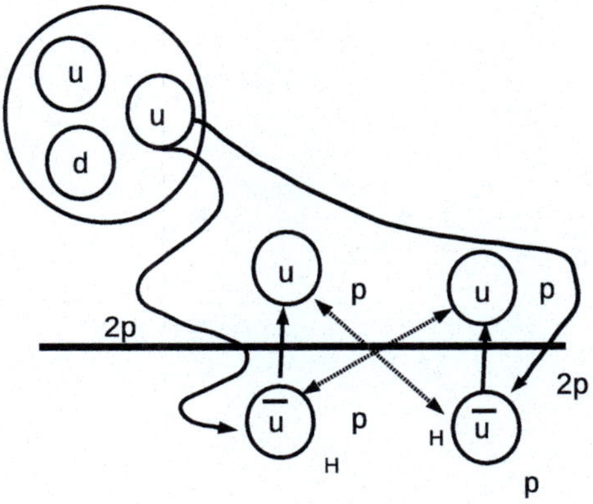

Fig. 7.4 Figure shows Higgs production from proton collisions

protons can then give W^+ and W^- bosons, respectively, which can combine to form Z boson which readily decays intro electron-positron pair.

7.4 Problems

1. Calculate the approximate (without spinor sum) differential cross section of electron-positron collision at 1 GeV each.
2. Calculate the approximate differential cross section of electron-positron collision at 1 GeV each to produce muon-antimuon pair.
3. Calculate the approximate differential cross-section of $\nu_e + \mu \rightarrow \nu_\mu + e$ collision at 1 GeV each.
4. Calculate the approximate differential cross section of $\nu_e + \mu \rightarrow \nu_e + \mu$ collision at 1 GeV each.
5. Calculate scattering cross section of electrons of protons when electron energy is 10 KeV.

References

1. R. Hofstadter, Electron scattering and nuclear structure. Rev. Mod. Phys. **28**, 214–254 (1956)
2. M. Thomson, *Modern particle Physics* (Cambridge University Press, 2013)
3. C. Amsler, Deep inelastic electron-proton scattering. Nucl. Particle Phys., 18-1–18-12 (2014)

The manufacturer's authorised representative in the EU is Springer
Nature Customer Service Centre GmbH, Europaplatz 3, 69115 Heidelberg,
Germany. If you have any concerns regarding our products, please
contact ProductSafety@springernature.com

Printed and bound by CPI Group (UK) Ltd, Croydon, CR0 4YY
30/09/2026
02225604-0005